LOGICS OF TIME AND COMPUTATION

CSLI
Lecture Notes
No. 7

LOGICS OF TIME AND COMPUTATION

Second Edition
Revised and Expanded

Robert Goldblatt

Printed in the United States

CIP data and other information appear at the end of the book

To my daughter Hannah

Preface to the First Edition

These notes are based on lectures, given at Stanford in the Spring Quarter of 1986, on modal logic, emphasising temporal and dynamic logics. The main aim of the course was to study some systems that have been found relevant recently to theoretical computer science.

Part One sets out the basic theory of normal modal and temporal propositional logics, covering the *canonical model* construction used for completeness proofs, and the *filtration* method of constructing finite models and proving decidability results and completeness theorems.

Part Two applies this theory to logics of discrete (integer), dense (rational), and continuous (real) time; to the temporal logic of *henceforth*, *next*, and *until*, as used in the study of concurrent programs; and to the propositional *dynamic* logic of regular programs.

Part Three is devoted to first-order dynamic logic, and focuses on the relationship between the computational process of assignment to a variable, and the syntactic process of substitution for a variable. A completeness theorem is obtained for a proof theory with an infinitary inference rule.

There is more material here than was covered in the course, partly because I have taken the opportunity to gather together a number of observations, new proofs of old theorems etc., that have occurred to me from time to time. Those familiar with the subject will observe, for instance, that in Part Two proofs of completeness for various logics of discrete and continuous time, and for the temporal logic of concurrency, as well as the discussion of Bull's theorem on normal extensions of S4.3, all differ from those that appear in the literature.

In order to make the notes effective for classroom use, I have deliberately presented much of the material in the form of exercises (especially in Part One). These exercises should therefore be treated as an integral part of the text.

Acknowledgements. My visit to Stanford took place during a period of sabbatical leave from the Victoria University of Wellington which was supported by both universities, and the Fulbright programme. I would like to thank Solomon Feferman and Jon Barwise for the facilities that were made available to me at that time. The CSLI provided generous access to its excellent computer-typesetting system, and the Center's Editor, Dikran Karagueuzian, was particularly helpful with technical advice and assistance in the preparation of the manuscript.

Preface to the Second Edition

The text for this edition has been increased by more than a third. Major additions are as follows.

- §7, originally concerned with incompleteness, now discusses a number of other metatheoretic topics, including first-order definability, (in)validity in canonical frames, failure of the finite model property, and the existence of undecidable logics with decidable axiomatisation.
- §9 now includes a study of the " branching time" system of *Computational Tree Logic*, due to Clarke and Emerson, which introduces connectives that formalise reasoning about behaviour along different branches of the tree of possible future states. Completeness and decidability are shown by the method of filtration in an adaptation of ideas due to Emerson and Halpern.
- In §10 dynamic logic is extended by the concurrency command $\alpha \cap \beta$, interpreted as "α and β executed in parallel". This is modelled by the use of "reachability relations", in which the outcome of a single execution is a set of terminal states, rather than a single state. This leads to a semantics for $[\alpha]$ and $<\alpha>$ which makes them independent (i.e. not interdefinable via negation). The resulting logic is shown to be finitely axiomatisable and decidable, by a new theory of canonical models and filtrations for reachability relations.

A significant conceptual change involves the definition of a "logic" (p. 16), which no longer includes the rule of Uniform Substitution. Logics satisfying this rule are called Uniform, and are discussed in detail on page 23. The change causes a number of minor adaptations throughout the text.

A notable technical improvement concerns the completeness proof for *S*4.3*Dum* in §8 (pp. 73–75). The original *Dum*-Lemma has been replaced by a direct proof that non-last clusters in the filtration are simple. This has resulted in some re-arrangement of the material concerning Bull's Theorem, and a simplification of the completeness theorem for the temporal logic of concurrency in §9 (pp. 95–96).

Other small changes include additional material about the Diodorean modality of spacetime (p. 45), and a rewriting of the basic filtration construction for dynamic logic (p. 114) using a uniform method of proving the first filtration condition that obviates the need to establish any standard-model conditions for the canonical model.

Reformatting the text has provided the opportunity to make numerous changes in style and expression, as well as to correct typos. I will be thankful for, if not pleased by, information about any further such errors.

rob@math.vuw.ac.nz

Contents

Part One

Propositional Modal Logic

1 | Syntax and Semantics

BNF

The notation of *Backus-Naur form* (BNF) will be used to define the syntax of the languages we will study. This involves specifying certain syntactic categories, and then giving recursive equations to show how the members of those categories are generated. The method can be illustrated by the syntax of standard propositional logic, which has one main category, that of the *formulae*. These are generated from some set of *atomic* formulae (or propositional variables), together with a constant $\perp$ (the *falsum*), by the connective $\rightarrow$ (implication). In BNF, this is expressed in one line as

$$<\text{formula}> ::= <\text{atomic formula}> \mid \perp \mid <\text{formula}>\rightarrow<\text{formula}>$$

The symbol ::= can be read "comprises", or "consists of", or simply "is". The vertical bar | is read "or". Thus the equation says that a formula is either an atomic formula, the falsum, or an implication between two formulae.

The definition becomes even more concise when we use individual letters for members of syntactic categories, in the usual way. Let Φ be a denumerable set of *atomic formulae*, with typical member denoted p. The set of all formulae generated from Φ will be denoted $Fma(\Phi)$, and its members denoted A, A_1, A', $B, \ldots$ etc. The presentation of syntax then becomes

Atomic formulae: $p \in \Phi$
Formulae: $A \in Fma(\Phi)$

$$A ::= p \mid \perp \mid A \rightarrow A$$

Technically, the recursive equation governs a non-deterministic rewriting procedure for generating formulae, in which any occurrence of the symbol to the left of the ::= sign can be replaced by any of the alternative expressions on the right side. Thus the two occurrences of A in the expression $A \rightarrow A$ may themselves be replaced by different expressions, and so stand for different formulae. In some BNF presentations, this is emphasised by using subscripts to distinguish different occurrences of a symbol. Then the above equation is given as

$$A ::= p \mid \bot \mid A_1 \to A_2.$$

Modal Formulae

The language of propositional modal logic requires one additional symbol, the "box" $\Box$. The BNF definition of the set of modal formulae generated by Φ is

Atomic formulae:	$p \in \Phi$
Formulae:	$A \in Fma(\Phi)$

$$A ::= p \mid \bot \mid A_1 \to A_2 \mid \Box A$$

Possible readings of $\Box A$

It is necessarily true that A.
It will always be true that A.
It ought to be that A.
It is known that A.
It is believed that A.
It is provable in Peano Arithmetic that A.
After the program terminates, A.

Other connectives

These are introduced by the usual abbreviations.

Negation:	$\neg A$	is	$A \to \bot$
Verum:	$\top$	is	$\neg\bot$
Disjunction:	$A_1 \vee A_2$	is	$(\neg A_1) \to A_2$
Conjunction:	$A_1 \wedge A_2$	is	$\neg(A_1 \to \neg A_2)$
Equivalence:	$A_1 \leftrightarrow A_2$	is	$(A_1 \to A_2) \wedge (A_2 \to A_1)$
"Diamond":	$\Diamond$	is	$\neg\Box\neg A$

Notational Convention

In the case that $n = 0$, the expression

$$B_0 \wedge \ldots \wedge B_{n-1} \to B$$

just denotes the formula B.

Exercises 1.1

(1) Decide what $\Diamond A$ means under each of the above readings of $\Box$.

(2) Which of the following should be regarded as true under the different readings of $\Box$?

$$\Box A \to A$$
$$\Box A \to \Box\Box A$$
$$\Diamond \top$$
$$\Box A \to \Diamond A$$
$$\Box A \vee \Box\neg A$$
$$\Box(A \to B) \to (\Box A \to \Box B)$$
$$\Diamond A \wedge \Diamond B \to \Diamond(A \wedge B)$$
$$\Box(\Box A \to A) \to \Box A$$

Subformulae

The finite set $Sf(A)$ of all subformulae of $A \in Fma(\Phi)$ is defined inductively by

$$Sf(p) = \{p\}$$
$$Sf(\bot) = \{\bot\}$$
$$Sf(A_1 \to A_2) = \{A_1 \to A_2\} \cup Sf(A_1) \cup Sf(A_2)$$
$$Sf(\Box A) = \{\Box A\} \cup Sf(A)$$

Schemata

We will often have occasion to refer to a *schema*, meaning a collection of formulae all having a common syntactic form. Thus, for example, by the schema

$$\Box A \to A$$

we mean the collection of formulae

$$\{\Box B \to B : B \in Fma(\Phi)\}.$$

Uniform Substitution

The notion of a schema can be made more precise by considering *uniform substitutions*, as follows.

Let A and B be any formulae, and p an atomic formula. By the *uniform substitution of B for p in A* we mean the procedure of replacing each and every occurrence of p in A by B. A formula A' is called a *substitution instance of A* if it arises by simultaneous uniform substitution for some of of the atomic formulae of A, i.e. if there exist some finitely many atomic formulae $p_1, \ldots, p_n$, and formulae $B_1, \ldots, B_n$, such that A' is the result of simultaneously uniformly substituting B_1 for p_1 in A, and B_2 for p_2 in A, and ..., and B_n for p_n in A. Let

$$\Sigma_A = \{A' : A' \text{ is a substitution instance of } A\}.$$

Then a schema may be defined as a set of formulae that is equal to Σ_A for some formula A.

For example, if A is the formula $\Box p \to p$, with p atomic, then Σ_A is what was described above as "the schema $\Box A \to A$".

Frames and Models

A *frame* is a pair $\mathcal{F} = (S, R)$, where S is a non-empty set, and R a binary relation on S: in symbols, $R \subseteq S \times S$.

A *Φ-model* on a frame is a triple $\mathcal{M} = (R, S, V)$, with $V : \Phi \to 2^S$. Hence V is a function assigning to each atomic formula $p \in \Phi$ a subset $V(p)$ of S. Informally, $V(p)$ is to be thought of as the set of points at which p is "true". Generally we drop the prefix Φ- in discussing models, provided the context is clear.

The relation "*A is true (holds) at point s in model $\mathcal{M}$*", denoted

$$\mathcal{M} \models_s A,$$

is defined inductively on the formation of $A \in Fma(\Phi)$ as follows.

$\mathcal{M} \models_s p$	iff	$s \in V(p)$
$\mathcal{M} \not\models_s \bot$		(i.e. not $\mathcal{M} \models_s \bot$)
$\mathcal{M} \models_s (A_1 \to A_2)$	iff	$\mathcal{M} \models_s A_1$ implies $\mathcal{M} \models_s A_2$
$\mathcal{M} \models_s \Box A$	iff	for all $t \in S$, sRt implies $\mathcal{M} \models_t A$

Exercises 1.2

(1) $\mathcal{M} \models_s \neg A$ iff $\mathcal{M} \not\models_s A$.
Work out the corresponding truth conditions for $A \wedge B$, $A \vee B$, $A \leftrightarrow B$.

(2) $\mathcal{M} \models_s \Diamond A$ iff there exists $t \in S$ with sRt and $\mathcal{M} \models_t A$.

Motivations

1. *Necessity.* Following the dictum of Leibnitz that a necessary truth is one that holds in all "possible worlds", S may be thought of as a set of such worlds, with sRt when t is a conceivable alternative to s, i.e. a world in which all the necessary truths of s are realised. $\Box A$ then means "A is necessarily true", while $\Diamond A$ means "A is possible", i.e. true in some conceivable world.

2. Different notions of necessity can be entertained. Thus *logical* necessity may be contrasted with *physical* necessity, the latter taking $\Box A$ to mean "A is a consequence of the laws of physics". Under this reading, sRt holds when t is a scientific alternative to s, i.e. a world in which all scientific laws of s are fulfilled. Hence in our world, $\Box(x < c)$ is true under the physical reading, where c is the velocity of light and x the velocity of a material body. On the other hand it is *logically* possible that $(x < c)$ is false.

3. In *deontic* logic, $\Box$ means "A ought to be true". sRt then means that t is a morally ideal alternative to s, a world in which all moral laws of s are obeyed. If s is the actual world, few would maintain that sRs under this interpretation. On the other hand, any world is a logical, and scientific, alternative to itself.

4. *Temporal Logic.* Here the members of S are taken to be moments of time. If sRt means "t is after (later than) s", then $\Box A$ means "*henceforth* A", i.e. "at all future times A", while $\Diamond A$ means "*eventually* (at some future time) A". Dually, if sRt means that t is before s, then $\Box$ means "*hitherto*", and so on. Natural time frames (S, R) for temporal logic are given by taking S as one of the number sets ω (natural numbers), $\mathbb{Z}$ (integers), $\mathbb{Q}$ (rationals), or $\mathbb{R}$ (reals), and R as one of the relations $<, \leq, >, \geq$. Another interesting possibility is to consider various orderings on the points of four-dimensional Minkowskian spacetime (cf. page 45, and Goldblatt [1980]), or even more general non-linear "branchings" in time (Rescher and Urquhart [1971]).

5. *Program states.* Reading $\Box$ as "after the program terminates", S is to be regarded as the set of possible *states* of a computation process, with sRt meaning that there is an execution of the program that starts in state s and terminates in state t. A non-deterministic program may admit more than one possible "outcome" t when started in s. Then $\Box A$ means "every terminating execution of the program brings about A", while $\Diamond A$ means that the program *enables* A, i.e. "there is some execution that terminates with A true".

   At the level of propositional logic, the notion of *state* is formally taken to be primitive, as in the theory of automata, Turing machines, etc. A natural concrete interpretation of the notion is possible in quantificational logic, as will be seen in Part Three.

Valuations and Tautologies

Given a Φ-model $\mathcal{M}$, and a fixed $s \in S$, define

$$V_s(p) = \begin{cases} \mathit{true} & \text{if } s \in V(p); \\ \mathit{false} & \text{otherwise.} \end{cases}$$

Then the function $V_s : \Phi \to \{true,\ false\}$ is a *valuation* of the atomic formulae, a notion familiar from propositional logic. Using the standard truth-tables for propositional connectives, V_s is extended to assign a truth-value to any formula not containing the symbol $\Box$.

Thus a model on a frame gives rise to a collection $\{V_s : s \in S\}$ of valuations of Φ, while, conversely, such a collection defines the model in which $V(p) = \{s : V_s(p) = true\}$.

A formula A is *quasi-atomic* if either it is atomic ($A \in \Phi$), or else it begins with a $\Box$, i.e. $A = \Box B$ for some B. If Φ^q is the set of all quasi-atomic formulae, then any formula A is constructible from members of $\Phi^q \cup \{\bot\}$ using the connective $\to$. Hence by using the truth-table for $\to$, any valuation

$$V : \Phi^q \to \{true,\ false\}$$

of the quasi-atomic formulae extends uniquely to a valuation

$$V : Fma(\Phi) \to \{true,\ false\}$$

of all formulae. A formula A is a *tautology* if $V(A) = true$ for every valuation V of its quasi-atomic subformulae.

Exercise 1.3

Any tautology is a substitution instance of a tautology of propositional logic (i.e. a $\Box$-free tautology).

Truth and Validity

Formula A is *true in model* $\mathcal{M}$, denoted $\mathcal{M} \models A$, if it is true at all points in $\mathcal{M}$, i.e. if

$$\mathcal{M} \models_s A \text{ for all } s \in S.$$

A is *valid in frame* $\mathcal{F} = (S, R)$, denoted $\mathcal{F} \models A$, if

$$\mathcal{M} \models A \text{ for all models } \mathcal{M} = (S, R, V) \text{ based on } \mathcal{F}.$$

If $\mathcal{C}$ is a class of models (respectively, frames), then A is *true* (respectively, *valid*) in $\mathcal{C}$, $\mathcal{C} \models A$, if A is true (respectively, valid) in all members of $\mathcal{C}$.

A schema will be said to be true in a model (respectively, valid in a frame) if all instances of the schema have that property. More generally, we will use the notations $\mathcal{M} \models \Gamma$ and $\mathcal{F} \models \Gamma$, where $\Gamma \subseteq Fma$, to mean that all members of Γ are true in $\mathcal{M}$, or valid in $\mathcal{F}$.

Exercises 1.4

(1) The following are true in all models, hence valid in all frames.

$$\Box\top$$
$$\Box(A \to B) \to (\Box A \to \Box B)$$
$$\Diamond(A \to B) \to (\Box A \to \Diamond B)$$
$$\Box(A \to B) \to (\Diamond A \to \Diamond B)$$
$$\Box(A \wedge B) \leftrightarrow (\Box A \wedge \Box B)$$
$$\Diamond(A \vee B) \leftrightarrow (\Diamond A \vee \Diamond B)$$

(2) Show that the following do not have the property of being valid in all frames.

$$\Box A \to A$$
$$\Box A \to \Box\Box A$$
$$\Box(A \to B) \to (\Box A \to \Diamond B)$$
$$\Diamond\top$$
$$\Diamond A \to \Box A$$
$$\Box(\Box A \to B) \vee \Box(\Box B \to A)$$
$$\Box(A \vee B) \to \Box A \vee \Box B$$
$$\Box(\Box A \to A) \to \Box A$$

(N.B. some *instances* of these schemata may be valid, e.g. when A is a tautology. What is required is to find a counterexample to validity of at least one instance of each schema.)

(3) Show that $\Diamond\top$ and the schema $\Box A \to \Diamond A$ have exactly the same models.

(4) Exhibit a frame in which $\Box\bot$ is valid.

(5) In any model $\mathcal{M}$,
 (i) if A is a tautology then $\mathcal{M} \models A$;
 (ii) if $\mathcal{M} \models A$ and $\mathcal{M} \models A \to B$, then $\mathcal{M} \models B$;
 (iii) if $\mathcal{M} \models A$ then $\mathcal{M} \models \Box A$.

(6) Items (i)-(iii) of the previous exercise hold if $\mathcal{M}$ is replaced by any frame $\mathcal{F}$.

Ancestral (Reflexive Transitive Closure)

Let $\mathcal{F} = (S, R)$ be a frame. Define on S the relations $R^n \subseteq S \times S$, for $n \geq 0$, and R^*, as follows.

$$sR^0t \quad \text{iff} \quad s = t$$
$$sR^{n+1}t \quad \text{iff} \quad \exists u(sR^nu \ \& \ uRt)$$

$$R^* \quad = \bigcup_{n \geq 0} R^n.$$

Exercises 1.5

(1) $R^1 = R$.

(2) sR^*t iff $\exists n \geq 0 \, \exists s_0, \ldots, \exists s_n \in S$ with $s_0 = s$, $s_n = t$, and for all $i < n$, $s_i R s_{i+1}$.

(3) R^* is reflexive and transitive.

(4) If T is any reflexive and transitive relation on S with $R \subseteq T$, then $R^* \subseteq T$. That is, R^* is the smallest reflexive and transitive relation on S that contains R.

(5) If $S \subseteq \mathbb{Z}$, and $R = \{(s, s+1) : s \in S\}$, what is R^*?

R^* is often known as the *ancestral* of R (from the case that R is the "parent of" relation). In view of exercise (4), it is also known as the *reflexive transitive closure* of R. The notion will play an important role in the logic of programs in Parts Two and Three.

Generated Submodels

If $\mathcal{M} = (S, R, V)$ and $t \in S$, then the *submodel of $\mathcal{M}$ generated by t* is

$$\mathcal{M}^t = (S^t, R^t, V^t),$$

where

$$\begin{aligned} S^t &= \{u \in S : tR^*u\} \\ R^t &= R \cap (S^t \times S^t) \\ V^t(p) &= V(p) \cap S^t. \end{aligned}$$

The structure $\mathcal{F}^t = (S^t, R^t)$ is the *subframe of $\mathcal{F} = (S, R)$ generated by t.*

Exercises 1.6

(1) If R is transitive, then $S^t = \{t\} \cup \{u : tRu\}$.

(2) S^t is the smallest subset X of S that contains t and is *closed under R*, in the sense that $u \in X$ and uRv implies $v \in X$.

To evaluate the truth of formula A at point t may require investigating the truth of certain subformulae B of A at various R-alternatives v of t. But then to determine the truth-value of B at v may require looking at alternatives of v. And so on. S^t comprises all points generated by this process. It is evident that evaluating truth at t will only involve points that are each obtainable from t by finitely many "R-alternations". This is embodied in the

Submodel Lemma 1.7. *If $A \in Fma(\Phi)$, then for any $u \in S^t$,*

$$\mathcal{M}^t \models_u A \quad \text{iff} \quad \mathcal{M} \models_u A.$$

Proof. By induction on the formation of A. The case $A = p \in \Phi$ follows from the definition of V^t, and the case $A = \bot$ is immediate. The inductive cases $A = (B \to D)$ and $A = \Box B$ are given as exercises.

Corollary 1.8.

(1) $\mathcal{M} \models A$ implies $\mathcal{M}^t \models A$.
(2) $\mathcal{M} \models A$ iff A is true in all generated submodels of $\mathcal{M}$.
(3) $\mathcal{F} \models A$ iff A is valid in all generated subframes of $\mathcal{F}$.

p-Morphisms

Let $\mathcal{M}_1 = (S_1, R_1, V_1)$ and $\mathcal{M}_2 = (S_2, R_2, V_2)$ be models, and $f : S_1 \to S_2$ a function satisfying

$$\begin{array}{lll} sR_1t & \text{implies} & f(s)R_2f(t); \\ f(s)R_2u & \text{implies} & \exists t(sR_1t \ \& \ f(t) = u); \\ s \in V_1(p) & \text{iff} & f(s) \in V_2(p). \end{array}$$

Then f is called a *p-morphism* from $\mathcal{M}_1$ to $\mathcal{M}_2$. A function satisfying the first two conditions is a *p-morphism from frame* (S_1, R_1) *to frame* (S_2, R_2).

p-Morphism Lemma 1.9. *If $A \in Fma(\Phi)$, then for any $s \in S_1$,*

$$\mathcal{M}_1 \models_s A \quad \text{iff} \quad \mathcal{M}_2 \models_{f(s)} A.$$

Proof. Exercise.

If there is a p-morphism $f : \mathcal{F}_1 \to \mathcal{F}_2$ that is *surjective* (onto), then frame $\mathcal{F}_2$ is called a *p-morphic image* of $\mathcal{F}_1$.

p-Morphism Lemma 1.10. *If $\mathcal{F}_2$ is a p-morphic image of $\mathcal{F}_1$, then for any formula A,*

$$\mathcal{F}_1 \models A \quad \text{implies} \quad \mathcal{F}_2 \models A.$$

Proof. Suppose A is false at some point t in some model $\mathcal{M}_2 = (\mathcal{F}_2, V_2)$ based on $\mathcal{F}_2$. Take a surjective p-morphism $f : S_1 \to S_2$ and define a model $\mathcal{M}_1 = (\mathcal{F}_1, V_1)$ by declaring

$$s \in V_1(p) \quad \text{iff} \quad f(s) \in V_2(p).$$

Then f is a p-morphism from $\mathcal{M}_1$ to $\mathcal{M}_2$. Choosing any s with $f(s) = t$, the first p-Morphism Lemma 1.9 gives A false at s in the model $\mathcal{M}_1$ based on $\mathcal{F}_1$.

Exercise 1.11

Let $\mathcal{F}_1 = (\{0,1\}, R)$ and $\mathcal{F}_2 = (\{0\}, R)$, where in each case R is the universal relation $S \times S$. Show that

$$\mathcal{F}_1 \models A \quad \text{implies} \quad \mathcal{F}_2 \models A;$$

$$(\omega, <) \models A \quad \text{implies} \quad \mathcal{F}_1 \models A.$$

The curious appellation "p-morphism" derives from an early use of the name "pseudo-epimorphism" in this context, and seems to have become entrenched in the literature.

Conditions on R

The following is a list of properties of a binary relation R that are defined by first-order sentences.

1. Reflexive: $\forall s(sRs)$
2. Symmetric: $\forall s \forall t(sRt \to tRs)$
3. Serial: $\forall s \exists t(sRt)$
4. Transitive: $\forall s \forall t \forall u(sRt \wedge tRu \to sRu)$
5. Euclidean: $\forall s \forall t \forall u(sRt \wedge sRu \to tRu)$
6. Partially functional: $\forall s \forall t \forall u(sRt \wedge sRu \to t = u)$
7. Functional: $\forall s \exists! t(sRt)$
8. Weakly dense: $\forall s \forall t(sRt \to \exists u(sRu \wedge uRt))$
9. Weakly connected: $\forall s \forall t \forall u(sRt \wedge sRu \to tRu \vee t = u \vee uRt)$
10. Weakly directed: $\forall s \forall t \forall u(sRt \wedge sRu \to \exists v(tRv \wedge uRv))$

Corresponding to this list is a list of schemata:

1. $\Box A \to A$
2. $A \to \Box \Diamond A$
3. $\Box A \to \Diamond A$
4. $\Box A \to \Box \Box A$
5. $\Diamond A \to \Box \Diamond A$
6. $\Diamond A \to \Box A$
7. $\Diamond A \leftrightarrow \Box A$
8. $\Box \Box A \to \Box A$
9. $\Box(A \wedge \Box A \to B) \vee \Box(B \wedge \Box B \to A)$
10. $\Diamond \Box A \to \Box \Diamond A$

Theorem 1.12. *Let $\mathcal{F} = (S, R)$ be a frame. Then for each of the properties 1-10, if R satisfies the property, then the corresponding schema is valid in $\mathcal{F}$.*

Proof. We illustrate with the case of transitivity. Suppose that R is transitive. Let $\mathcal{M}$ be any model on $\mathcal{F}$. To show that

$$\mathcal{M} \models \Box A \to \Box\Box A,$$

take any s in $\mathcal{M}$ with $\mathcal{M} \models_s \Box A$. We have to prove

$$\mathcal{M} \models_s \Box\Box A,$$

which means

$$sRt \quad \text{implies} \quad \mathcal{M} \models_t \Box A,$$

or, in other words,

$$sRt \quad \text{implies} \quad (tRu \quad \text{implies} \quad \mathcal{M} \models_u A).$$

So, suppose sRt. Then if tRu, we have sRu by transitivity, so $\mathcal{M} \models_u A$, since $\mathcal{M} \models_s \Box A$ by hypothesis.

The other cases are left as exercises.

Theorem 1.13. *If a frame $\mathcal{F} = (S, R)$ validates any one of the schemata 1-10, then R satisfies the corresponding property.*

Proof. Take the case of schema 10. To show R is weakly directed, suppose sRt and sRu. Let $\mathcal{M}$ be any model on $\mathcal{F}$ in which $V(p) = \{v : uRv\}$. Then by definition,

$$uRv \quad \text{implies} \quad \mathcal{M} \models_v p,$$

so $\mathcal{M} \models_u \Box p$, and hence, as sRu, $\mathcal{M} \models_s \Diamond\Box p$. But then as schema 10 is valid in $\mathcal{F}$, $\mathcal{M} \models_s \Box\Diamond p$, so as sRt, $\mathcal{M} \models_t \Diamond p$. This implies that there exists a v with tRv and $\mathcal{M} \models_v p$, i.e. $v \in V(p)$, so uRv as desired.

Next, the case of schema 8. Suppose sRt. Let $\mathcal{M}$ be a model on $\mathcal{F}$ with $V(p) = \{v : t \neq v\}$. Then $\mathcal{M} \not\models_t p$, so $\mathcal{M} \not\models_s \Box p$. Hence by validity of schema 8, $\mathcal{M} \not\models_s \Box\Box p$, so there exists a u with sRu and $\mathcal{M} \not\models_u \Box p$. Then for some v, uRv and $\mathcal{M} \not\models_v p$, i.e. $v = t$, so that uRt, as needed to show that R is weakly dense.

Exercises 1.14

(1) Complete the proofs of Theorems 1.12 and 1.13.

(2) Give a property of R that is necessary and sufficient for $\mathcal{F}$ to validate the schema $A \to \Box A$. Do the same for $\Box\bot$.

First-Order Definability

Theorems 1.12 and 1.13 go a long way toward explaining the great success that the relational semantics enjoyed upon its introduction by Kripke [1963]. Frames are much easier to deal with than the modelling structures (Boolean algebras with a unary operator) that had been available hitherto, and many modal schemata were shown to have their frames characterised by simple first-order properties of R. For a time it seemed that propositional modal logic corresponded in strength to first-order logic, but that proved not to be so. Here are a couple of illustrations.

(1) The schema

$$W: \qquad \Box(\Box A \to A) \to \Box A$$

is valid in frame (S, R) iff

(i) R is transitive, and

(ii) there are no sequences $s_0, \ldots, s_n, \ldots$ in S with $s_n R s_{n+1}$ for all $n \geq 0$.

(for a proof cf. Boolos [1979], p.82). Now it can be shown by the Compactness Theorem of first-order logic that there exists a frame satisfying (i) and (ii) that is elementarily equivalent to (i.e. satisfies the same first-order sentences as) a frame in which (ii) fails. Hence there can be no set of first-order sentences that defines the class of frames of this schema.

(2) The class of frames of the so-called *McKinsey schema*

$$M: \qquad \Box\Diamond A \to \Diamond\Box A$$

is not defined by any set of first-order sentences (Goldblatt [1975], van Benthem [1975]).

(Both of the above schemata will figure in the discussion of incompleteness in §7, where there is also a further consideration of the question of first-order definability.)

Subsequent investigations demonstrated that propositional modal logic corresponds to a fragment of second-order logic (Thomason [1975]).

Undefinable conditions

There are some naturally occurring properties of a binary relation R that do not correspond to the validity of any modal schema. One such is *irreflexivity*, i.e. $\forall s \neg(sRs)$. To see this, observe that the class of all frames validating a given schema is closed under p-morphic images (1.10), but the class of irreflexive frames is not so closed. For instance, it contains $(\omega, <)$, but not its p-morphic image $(\{0\}, \{\langle 0, 0\rangle\})$ (cf. Exercise 1.11).

Exercise 1.15

Show that neither of the following conditions correspond to any modal schema.

Antisymmetry: $\forall s \forall t(sRt \wedge tRs \rightarrow s = t)$,
Asymmetry: $\forall s \forall t(sRt \rightarrow \neg tRs)$.

Historical Note

The concepts of *necessity* and *possibility* have been studied by philosophers throughout history, notably by Aristotle, and in the middle ages. The contemporary symbolic analysis of modality is generally considered to have originated in the work of C. I. Lewis early this century (cf. Lewis and Langford [1932]). Lewis was concerned with a notion of *strict* implication. He defined "*A strictly implies B*" as $\mathrm{I}(A \wedge \neg B)$, where I is a primitive *impossibility* operator (later he expressed this as $\neg\Diamond(A \wedge \neg B)$, where $\Diamond$ expresses possibility). He defined a series of systems, which he called $S1$ to $S5$, based directly on axioms for strict implication. The standard procedure nowadays is to adjoin axioms and rules for $\Box$, or $\Diamond$, to the usual presentation of propositional logic. This approach to modal logic was first used in a paper by Gödel [1933]. The model theory described in this section is due to Kripke [1959, 1963].

To learn about the history of modal logic, the reader should first consult the interesting Historical Introduction to Lemmon [1977], where further references may be found.

2 | Proof Theory

Logics

Given a language based on a countable set Φ of atomic formulae, a *logic* is defined to be any set $\Lambda \subseteq Fma(\Phi)$ such that

- Λ includes all tautologies, and
- Λ is closed under the rule of *Detachment*, i.e.,
  if A, $A \to B \in \Lambda$ then $B \in \Lambda$.

Examples of Logics

(1) $PL = \{A \in Fma(\Phi) : A \text{ is a tautology }\}$.

(2) For any class $\mathcal{C}$ of models, or of frames (including the cases $\mathcal{C} = \{\mathcal{M}\}$ and $\mathcal{C} = \{\mathcal{F}\}$),

$$\Lambda_{\mathcal{C}} = \{A : \mathcal{C} \models A\}$$

is a logic.

(3) $Fma(\Phi)$ itself is a logic.

(4) If $\{\Lambda_i : i \in I\}$ is a collection of logics, then their intersection

$$\bigcap_{i \in I} \Lambda_i$$

is a logic. Thus for any $\Gamma \subseteq Fma(\Phi)$ there is a *smallest* logic containing Γ, namely the intersection of the collection

$$\{\Lambda : \Lambda \text{ is a logic and } \Gamma \subseteq \Lambda\}.$$

Note that PL is the *smallest* logic, and $Fma(\Phi)$ the *largest*, in the sense that for any logic Λ,

$$PL \subseteq \Lambda \subseteq Fma(\Phi).$$

Tautological Consequence

A formula A is a *tautological consequence* of formulae $A_1, \ldots, A_n$ if A is assigned *true* by every valuation that assigns *true* to all of $A_1, \ldots, A_n$. In particular, a tautological consequence of the empty set of formulae is the same thing as a tautology.

Exercise 2.1

Show that any logic Λ is closed under tautological consequence, i.e. if $A_1, \ldots, A_n \in \Lambda$, then any tautological consequence of $A_1, \ldots, A_n$ belongs to Λ.

Instead of defining a logic Λ to include all tautologies, it would suffice to include a set of schemata from which all tautologies can be derived by Detachment, e.g. the schemata

$$A \to (B \to A)$$
$$A \to (B \to D) \to ((A \to B) \to (A \to D))$$
$$\neg\neg A \to A.$$

Theorems

The members of a logic are called its *theorems.* We write $\vdash_\Lambda A$ to mean that A is a Λ-theorem, i.e.,

$$\vdash_\Lambda A \quad \text{iff} \quad A \in \Lambda.$$

Soundness and Completeness

Let $\mathcal{C}$ be a class of frames, or of models. Then logic Λ is *sound with respect to* $\mathcal{C}$ if for all formulae A,

$$\vdash_\Lambda A \quad \text{implies} \quad \mathcal{C} \models A.$$

Λ is *complete with respect to* $\mathcal{C}$, if, for any A,

$$\mathcal{C} \models A \quad \text{implies} \quad \vdash_\Lambda A.$$

Λ is *determined by* $\mathcal{C}$ if it is both sound and complete with respect to $\mathcal{C}$.

Deducibility and Consistency

If $\Gamma \cup \{A\} \subseteq Fma(\Phi)$, then A is *Λ-deducible from* Γ, denoted $\Gamma \vdash_\Lambda A$, if there exist $B_0, \ldots, B_{n-1} \in \Gamma$ such that

$$\vdash_\Lambda B_0 \to (B_1 \to (\cdots \to (B_{n-1} \to A)\cdots))$$

(in the case $n = 0$, this means that $\vdash_\Lambda A$). We write $\Gamma \nvdash_\Lambda A$ when A is not Λ-deducible from Γ.

A set $\Gamma \subseteq Fma(\Phi)$ is *Λ-consistent* if $\Gamma \nvdash_\Lambda \bot$. A formula A is Λ-consistent if the set $\{A\}$ is.

Exercises 2.2

(1) $\vdash_\Lambda A$ iff $\emptyset \vdash_\Lambda A$.

(2) If $\vdash_\Lambda A$ then $\Gamma \vdash_\Lambda A$.

(3) If $\Lambda \subseteq \Lambda'$, then $\Gamma \vdash_\Lambda A$ implies $\Gamma \vdash_{\Lambda'} A$.

(4) If $A \in \Gamma$ then $\Gamma \vdash_\Lambda A$.

(5) If $\Gamma \subseteq \Delta$ and $\Gamma \vdash_\Lambda A$, then $\Delta \vdash_\Lambda A$.

(6) If $\Gamma \vdash_\Lambda A$ and $\{A\} \vdash_\Lambda B$, then $\Gamma \vdash_\Lambda B$.

(7) *Detachment:* If $\Gamma \vdash_\Lambda A$ and $\Gamma \vdash_\Lambda A \to B$, then $\Gamma \vdash_\Lambda B$.

(8) *Deduction Theorem:* $\Gamma \cup \{A\} \vdash_\Lambda B$ iff $\Gamma \vdash_\Lambda A \to B$.

(9) $\Gamma \vdash_\Lambda A$ iff there exists a finite sequence $A_0, \ldots, A_m = A$ such that for all $i \leq m$, either $A_i \in \Gamma \cup \Lambda$, or else $A_k = (A_j \to A_i)$ for some $j, k < i$ (i.e. A_i follows from A_j and A_k by Detachment).

(10) $\{A : \Gamma \vdash_\Lambda A\}$ is the smallest logic containing $\Gamma \cup \Lambda$.

(11) *Soundness:* If $\mathcal{M} \models_s \Gamma \cup \Lambda$ and $\Gamma \vdash_\Lambda A$, then $\mathcal{M} \models_s A$.

(12) If $\Gamma \subseteq \Lambda$, then Γ is Λ-consistent if, and only if, $\Lambda \neq Fma(\Phi)$.

(13) Γ is Λ-consistent iff there exists a formula A with $\Gamma \nvdash_\Lambda A$.

(14) Γ is Λ-consistent iff there is no formula A having both $\Gamma \vdash_\Lambda A$ and $\Gamma \vdash_\Lambda \neg A$.

(15) $\Gamma \vdash_\Lambda A$ iff $\Gamma \cup \{\neg A\}$ is not Λ-consistent.

(16) $\Gamma \cup \{A\}$ is Λ-consistent iff $\Gamma \nvdash_\Lambda \neg A$.

(17) If Γ is Λ-consistent, then for any formula A, at least one of $\Gamma \cup \{A\}$ and $\Gamma \cup \{\neg A\}$ is Λ-consistent.

Maximal Sets

Let $\mathcal{M} = (S, R, V)$ be a model of a logic Λ, i.e. $\mathcal{M} \models \Lambda$. Associate with each $s \in S$ the set

$$\Gamma_s = \{A \in Fma(\Phi) : \mathcal{M} \models_s A\}.$$

Then Γ_s is Λ-consistent (why?), and moreover, for each formula A, one of A and $\neg A$ is in Γ_s.

In the next section we will be building models for certain logics. Since we have only a syntactic structure, namely Λ, to begin with, we will have to use syntactic entities, such as formulae or sets of formulae, as the points of our models. It turns out that the way to proceed is to use sets of formulae that enjoy the properties possessed by those sets Γ_s naturally associated with points of a given Λ-model.

A set $\Gamma \subseteq Fma(\Phi)$ is defined to be Λ-*maximal* if

- Γ is Λ-consistent, and
- for any $A \in Fma(\Phi)$, either $A \in \Gamma$ or $\neg A \in \Gamma$.

We define

$$S^\Lambda = \{\Gamma : \Gamma \text{ is } \Lambda\text{-maximal}\}.$$

Exercises 2.3

Suppose Γ is Λ-maximal.

(1) $\Gamma \vdash_\Lambda A$ implies $A \in \Gamma$.

(2) If $A \notin \Gamma$, then $\Gamma \cup \{A\}$ is not Λ-consistent. Hence if $\Gamma \subseteq \Delta$ and Δ is Λ-consistent, then $\Gamma = \Delta$ (this explains the use of the adjective "maximal").

(3) For any formula A, *exactly one* of A and $\neg A$ belongs to Γ, i.e.,

$$\neg A \in \Gamma \quad \text{iff} \quad A \notin \Gamma.$$

(4) $\Lambda \subseteq \Gamma$.

(5) $\bot \notin \Gamma$.

(6) $(A \to B) \in \Gamma$ iff ($A \in \Gamma$ implies $B \in \Gamma$).

(7) $A \wedge B \in \Gamma$ iff $A, B \in \Gamma$.

(8) $A \vee B \in \Gamma$ iff $A \in \Gamma$ or $B \in \Gamma$.

(9) $(A \leftrightarrow B) \in \Gamma$ iff ($A \in \Gamma$ iff $B \in \Gamma$).

Existence of Maximal Sets

We have yet to show that $S^\Lambda \neq \emptyset$, i.e. that there are any Λ-maximal sets. To see this, let

$$A_0, A_1, \ldots, A_n, \ldots\ldots$$

be an enumeration of the set $Fma(\Phi)$ (such an enumeration exists, since Φ is countable). Then if Γ is any Λ-consistent set, define

$$\Delta_0 = \Gamma$$

$$\Delta_{n+1} = \begin{cases} \Delta_n \cup \{A_n\}, & \text{if } \Delta_n \vdash_\Lambda A_n; \\ \Delta_n \cup \{\neg A_n\}, & \text{otherwise.} \end{cases}$$

$$\Delta = \textstyle\bigcup_{n \geq 0} \Delta_n.$$

By construction, at least one of A_n and $\neg A_n$ is in Δ, for all n.

Exercises 2.4

(1) Δ_n is Λ-consistent, for all n.
(2) Exactly one of A and $\neg A$ is in Δ, for all formulae A.
(3) If $\Delta \vdash_\Lambda B$, then $B \in \Delta$.

It follows from these exercises that Δ is Λ-consistent. For, if $\Delta \vdash \bot$, then $\Delta_n \vdash \bot$ for some n, contrary to the consistency of Δ_n. Thus we have established

Lindenbaum's Lemma 2.5. *Every Λ-consistent set of formulae is contained in a Λ-maximal set.*

Corollary 2.6.

(1) $\{A : \Gamma \vdash_\Lambda A\} = \bigcap\{\Delta \in S^\Lambda : \Gamma \subseteq \Delta\}$,
i.e. $\Gamma \vdash_\Lambda A$ iff A belongs to every Λ-maximal set that includes Γ.
(2) $\Lambda = \bigcap\{\Delta : \Delta \in S^\Lambda\}$,
i.e. $\vdash_\Lambda A$ iff A belongs to every Λ-maximal set.

Proof. We prove only the deeper part of (1). If $\Gamma \nvdash_\Lambda A$, then $\Gamma \cup \{\neg A\}$ is Λ-consistent (2.2(15)), so for some $\Delta \in S^\Lambda$, $\Gamma \cup \{\neg A\} \subseteq \Delta$. Then Δ includes Γ but does not contain A, since it contains $\neg A$ and is Λ-consistent.

Normal Logics

A logic Λ is *normal* if it contains the schema

$$K: \quad \Box(A \to B) \to (\Box A \to \Box B),$$

and is closed under the rule of *Necessitation*, i.e.,

$$\text{if } \vdash_\Lambda A, \text{ then } \vdash_\Lambda \Box A.$$

Examples of Normal Logics

(1) For any class $\mathcal{C}$ of models, or of frames,

$$\Lambda_{\mathcal{C}} = \{A : \mathcal{C} \models A\}$$

is a normal logic.

(2) If $\{\Lambda_i : i \in I\}$ is a collection of normal logics, then

$$\bigcap\{\Lambda_i : i \in I\}$$

is normal. In particular,

$$K = \bigcap\{\Lambda : \Lambda \text{ is a normal logic}\}$$

is the *smallest* normal logic. The letter K here is in honour of Kripke.

Example 1 shows that any logic determined by relational models or frames is normal, and so this is the type of logic we will be dealing with throughout.

Exercises 2.7

(1) If Λ is a normal logic, show the following.

$$\vdash_\Lambda A \to B \quad \text{implies} \quad \vdash_\Lambda \Box A \to \Box B \text{ and } \vdash_\Lambda \Diamond A \to \Diamond B.$$
$$\vdash_\Lambda A \leftrightarrow B \quad \text{implies} \quad \vdash_\Lambda \Box A \leftrightarrow \Box B \text{ and } \vdash_\Lambda \Diamond A \leftrightarrow \Diamond B.$$
$$\vdash_\Lambda \Diamond\neg A \leftrightarrow \neg\Box A.$$
$$\vdash_\Lambda \Box A \wedge \Box B \leftrightarrow \Box(A \wedge B).$$
$$\vdash_\Lambda \Diamond(A \vee B) \leftrightarrow \Diamond A \vee \Diamond B.$$
$$\vdash_\Lambda \Box A \vee \Box B \to \Box(A \vee B).$$
$$\vdash_\Lambda \Diamond(A \wedge B) \to \Diamond A \wedge \Diamond B.$$

(2) A logic Λ is normal iff for all $n \geq 0$,

$$\begin{array}{ll} \text{if} & \vdash_\Lambda A_0 \wedge \ldots \wedge A_{n-1} \to A, \\ \text{then} & \vdash_\Lambda \Box A_0 \wedge \ldots \wedge \Box A_{n-1} \to \Box A. \end{array}$$

(Note: when $n = 0$, this is just the rule of Necessitation.)

(3) A logic Λ is normal iff it satisfies the following three conditions.

$$\vdash_\Lambda \Box\top,$$
$$\vdash_\Lambda \Box A \wedge \Box B \to \Box(A \wedge B),$$
$$\vdash_\Lambda A \to B \quad \text{implies} \quad \vdash_\Lambda \Box A \to \Box B.$$

(4) If Λ is normal, then

$$\Gamma \vdash_\Lambda A \quad \text{implies} \quad \{\Box B : B \in \Gamma\} \vdash_\Lambda \Box A.$$

(5) If a normal logic contains the schema

$$\Diamond A \to \Box A,$$

then it contains the schemata

$$\Box(A \vee B) \leftrightarrow (\Box A \vee \Box B),$$
$$(\Box A \to \Box B) \leftrightarrow \Box(A \to B).$$

(6) $\vdash_K A$ iff there is a sequence $A_0, \ldots, A_m = A$ such that for all $i \leq m$, either A_i is a tautology or an instance of K, or $A_k = (A_j \to A_i)$ for some $j, k < i$, or $A_i = \Box A_j$ for some $j < i$.

Some Standard Logics

It has become customary to use the notation

$$K\Sigma_1 \ldots \Sigma_n$$

to refer to the smallest normal logic containing the schemata $\Sigma_1, \ldots, \Sigma_n$. Set-theoretically this logic is defined as

$$\bigcap\{\Lambda : \Lambda \text{ is normal and } \Sigma_1 \cup \ldots \cup \Sigma_n \subseteq \Lambda\}.$$

Historical names for some well-known schemata are

D:	$\Box A \to \Diamond A$
T:	$\Box A \to A$
4:	$\Box A \to \Box\Box A$
B:	$A \to \Box\Diamond A$
5:	$\Diamond A \to \Box\Diamond A$
L:	$\Box(A \wedge \Box A \to B) \vee \Box(B \wedge \Box B \to A)$
W:	$\Box(\Box A \to A) \to \Box A$

Names of some well-known logics are

$$S4 = KT4$$
$$S5 = KT4B$$
$$G = KW$$
$$K4.3 = K4L$$
$$S4.3 = KT4L$$

Exercises 2.8

(1) A is a theorem of $K\Sigma_1 \ldots \Sigma_n$ iff there is a sequence $A_0, \ldots, A_m = A$ such that for all $i \leq m$, either A_i is a tautology, an instance of schema K, or an instance of some Σ_i, or else $A_k = (A_j \to A_i)$ for some $j, k < i$, or else $A_i = \Box A_j$ for some $j < i$.

(2) KD is the smallest normal logic containing the formula $\Diamond\top$.

(3) $KB4 = KB5$.

(4) $S5 = KDB4 = KDB5 = KT5$.

(5) In the definition of $S4.3$, the schema L can be simplified to

$$\Box(\Box A \to B) \vee \Box(\Box B \to A).$$

(6) $K4 \subseteq G$, i.e. $\vdash_{KW} \Box A \to \Box\Box A$ (cf. Boolos [1979], p.30).

Uniform Logics

A logic Λ is *uniform* if it is closed under Uniform Substitution, i.e.,

if $A \in \Lambda$, then $A' \in \Lambda$ whenever A' is a substitution instance of A

(cf. page 5 for the definition of "substitution instance").

Exercises 2.9

(1) Λ is uniform iff

$$A \in \Lambda \quad \text{implies} \quad \Sigma_A \subseteq \Lambda,$$

where Σ_A is the schema defined by A (cf. page 6).

(2) If Λ is uniform, and $\Lambda \cap \Phi \neq \emptyset$, then Λ is not consistent.

The logic $\Lambda_{\mathcal{M}} = \{A : \mathcal{M} \models A\}$ determined by a model $\mathcal{M}$ need not be uniform. For instance, if $V(p) = S$ in $\mathcal{M}$, then $\Lambda_{\mathcal{M}}$ will contain the atomic formula p, but not its substitution instance $\bot$. However, most of the logics we will encounter are uniform, including any logic determined by a frame, or a class of frames, and any of the form $K\Sigma_1 \dots \Sigma_n$, where the Σ_i are schemata. These results are covered as follows.

Exercises 2.10

(1) Associate with each atomic formula p a formula B_p. Then if $\mathcal{M} = (S, R, V)$ is a model on a frame $\mathcal{F} = (S, R)$, define a new model $\mathcal{M}' = (S, R, V')$ on $\mathcal{F}$ by putting

$$V'(p) = \{s \in S : \mathcal{M} \models_s B_p\}.$$

Prove that for any formula A, and any $s \in S$,

$$\mathcal{M}' \models_s A \quad \text{iff} \quad \mathcal{M} \models_s A',$$

where A' is the result of uniformly substituting B_p for each atomic formula p in A.

(2) Deduce from Exercise (1) that for any frame $\mathcal{F}$, the normal logic $\{A : \mathcal{F} \models A\}$ is uniform.

(3) Associate with each atomic formula p a formula B_p, and, as in Exercise (1), for each formula A, let A' be the result of uniformly substituting B_p for each atomic p in A.

Let $\Sigma_1, \dots, \Sigma_n$ be a list of schemata, and $A_0, \dots, A_m$ a sequence of formulae fulfilling the description given in Exercise 2.8(1). Show that the sequence $A'_0, \dots, A'_m$ also fulfills this description with A'_i in place of A_i.

(4) Deduce from Exercise (3) that any logic of the form $K\Sigma_1 \dots \Sigma_n$ is uniform.

3 | Canonical Models and Completeness

The *canonical model* of a consistent normal logic Λ is the structure

$$\mathcal{M}^\Lambda = (S^\Lambda, R^\Lambda, V^\Lambda),$$

where

$$S^\Lambda = \{s \subseteq Fma(\Phi) : s \text{ is } \Lambda\text{-maximal}\},$$
$$sR^\Lambda t \quad \text{iff} \quad \{A \in Fma(\Phi) : \Box A \in s\} \subseteq t,$$
$$V^\Lambda(p) = \{s \in S^\Lambda : p \in s\}.$$

With regard to the definition of R^Λ, recall the intuitive interpretation of sRt as meaning that t is a conceivable alternative to s, a world in which all necessary truths of s are realised.

The *canonical frame* for Λ is $\mathcal{F}^\Lambda = (S^\Lambda, R^\Lambda)$. (Note that if Λ is not consistent (i.e. $\vdash_\Lambda \bot$, and hence $\Lambda = Fma$), then there are no Λ-maximal sets, so $\mathcal{M}^\Lambda$ and $\mathcal{F}^\Lambda$ do not exist.)

Exercise 3.1

$sR^\Lambda t$ iff $\{\neg\Box A : A \notin t\} \subseteq s$ iff $\{\Diamond A : A \in t\} \subseteq s$.

Theorem 3.2. *For any $s \in S^\Lambda$, and any $B \in Fma(\Phi)$,*

$$\Box B \in s \quad \text{iff} \quad \text{for all } t \in S^\Lambda, sR^\Lambda t \text{ implies } B \in t.$$

Proof. We give the "if" direction only. Suppose that for all $t \in S^\Lambda$,

$$sR^\Lambda t \quad \text{implies} \quad B \in t,$$

i.e.,

$$\{A : \Box A \in s\} \subseteq t \quad \text{implies} \quad B \in t.$$

Then by Corollary 2.6(1),

$$\{A : \Box A \in s\} \vdash_\Lambda B,$$

so by Exercise 2.7(4),

$$\{\Box A : \Box A \in s\} \vdash_\Lambda \Box B,$$

and hence by Exercise 2.2(5),

$$s \vdash_\Lambda \Box B.$$

But s is Λ-deductively closed (Ex. 2.3(1)), and so $\Box B \in s$ as desired.

Truth Lemma 3.3. *Let $A \in Fma(\Phi)$. Then for any $s \in S^{\Lambda}$,*

$$\mathcal{M}^{\Lambda} \models_s A \quad \text{iff} \quad A \in s.$$

Proof. By induction on the formation of A. The case $A = p \in \Phi$ is given by the definition of V^{Λ}, while the case $A = \bot$, and the inductive case $A = (B \to C)$, follow from Exercises 2.3(5) and 2.3(6). For the case $A = \Box B$, assume inductively that the result holds for B, and apply Theorem 3.2.

Corollary 3.4. *$\mathcal{M}^{\Lambda}$ determines Λ, i.e. for all formulae A,*

$$\mathcal{M}^{\Lambda} \models A \quad \text{iff} \quad \vdash_{\Lambda} A.$$

Proof. By Corollary 2.6, $\vdash_{\Lambda} A$ iff A belongs to all members of S^{Λ}.

Note that Corollary 3.4 implies that Λ is *complete* with respect to the frame $\mathcal{F}^{\Lambda}$:

$$\mathcal{F}^{\Lambda} \models A \quad \text{implies} \quad \vdash_{\Lambda} A.$$

Λ need not however be *sound* with respect to $\mathcal{F}^{\Lambda}$, i.e. it may be that $\mathcal{F}^{\Lambda} \not\models \Lambda$. Indeed there are some logics that are not determined by any class of frames. §7 will discuss examples.

Theorem 3.5. (Determination of K). *$\vdash_K A$ if, and only if, A is valid in all frames.*

Proof.

Soundness: For any frame $\mathcal{F}$, $\Lambda_{\mathcal{F}} = \{B : \mathcal{F} \models B\}$ is a normal logic, so $K \subseteq \Lambda_{\mathcal{F}}$, i.e. $\vdash_K A$ implies $\mathcal{F} \models A$.

Completeness: if $\nvdash_K A$, then by Corollary 3.4, A is false in $\mathcal{M}^K$, and so is not valid in the frame $\mathcal{F}^K$.

Completeness Theorems

In order to show that a logic Λ is complete with respect to some class of models, or of frames, defined by certain conditions, it suffices to show that $\mathcal{M}^{\Lambda}$, or $\mathcal{F}^{\Lambda}$, satisfies those conditions. The great power of this approach resides in the fact that the proof-theoretic properties of Λ have an impact on the properties of the relation R^{Λ}. To give some examples of this, recall the first-order properties 1-10 of R, and their corresponding modal schemata, listed on page 12.

Theorem 3.6. *If a normal logic Λ contains any one of the schemata 1-10, then R^{Λ} satisfies the corresponding first-order condition.*

Proof. Generally, the proof for a universal condition, like transitivity, is a relatively straightforward application of the definitions, while cases which involve existential assertions, such as weak density, require a deeper construction. We illustrate with these two properties.

Transitivity. Suppose Λ contains the schema

$$\Box A \to \Box\Box A.$$

Then all members of S^Λ contain all instances of this schema. Hence if $sR^\Lambda t$ and $tR^\Lambda u$, $\Box A \in s$ implies $\Box\Box A \in s$, so $\Box A \in t$ as $sR^\Lambda t$, and then $A \in u$ as $tR^\Lambda u$. This proves

$$\{A : \Box A \in s\} \subseteq u,$$

i.e. $sR^\Lambda u$ as desired.

Weak density. Suppose Λ contains the schema

$$\Box\Box A \to \Box A.$$

Assume $sR^\Lambda t$. We want to find some $u \in S^\Lambda$ such that $sR^\Lambda u$, i.e. $\{A : \Box A \in s\} \subseteq u$, and $uR^\Lambda t$, which is equivalent to $\{\neg\Box B : B \notin t\} \subseteq u$ (cf. Exercise 3.1). Therefore it suffices to show that the set

$$u_0 = \{A : \Box A \in s\} \cup \{\neg\Box B : B \notin t\}$$

is Λ-consistent. For then by Lindenbaum's Lemma 2.5, there will be some $u \in S^\Lambda$ with $u_0 \subseteq u$ as desired.

Suppose then that u_0 is not Λ-consistent. Then there is a Λ-theorem

$$\vdash_\Lambda A_1 \wedge \ldots \wedge A_m \wedge \neg\Box B_1 \wedge \ldots \wedge \neg\Box B_n \to \bot, \qquad \text{(i)}$$

for some $m, n \geq 0$, with each $\Box A_i$ in s, and each B_j not in t. Let $B = (B_1 \vee \ldots \vee B_n)$. Then since

$$\vdash_\Lambda \Box B_1 \vee \ldots \vee \Box B_n \to \Box B \qquad \text{(ii)}$$

(cf. Ex. 2.7(1)), it follows from (i) and (ii) by tautological consequence that

$$\vdash_\Lambda A_1 \wedge \ldots \wedge A_m \to \Box B$$

and so by Exercise 2.7(2),

$$\vdash_\Lambda \Box A_1 \wedge \ldots \wedge \Box A_m \to \Box\Box B.$$

As each $\Box A_i$ is in s, this implies $\Box\Box B \in s$. But by hypothesis,

$$(\Box\Box B \to \Box B) \in s,$$

hence $\Box B \in s$, giving $B \in t$ as $sR^\Lambda t$. But this implies that for some j, B_j is in t (2.3(8)), which is a contradiction.

Thus the hypothesis that u_0 is Λ-inconsistent must be false.

Exercise 3.7

Complete the proof of Theorem 3.6.

The next Theorem and Exercises exemplify the way in which canonical models are used to prove completeness and determination results. The particular logics concerned were defined on page 22.

Theorem 3.8. *$S4$ is determined by the class of reflexive and transitive frames.*

Proof.

Soundness. If the relation R of frame $\mathcal{F}$ is reflexive and transitive, then the normal logic

$$\Lambda_{\mathcal{F}} = \{A : \mathcal{F} \models A\}$$

contains the schemata T and 4, and so contains $KT4 = S4$, i.e. $\vdash_{S4} A$ implies $\mathcal{F} \models A$.

Completeness. By the schemata T and 4, the canonical $S4$-frame is reflexive and transitive (Theorem 3.6). Hence if A is valid in all reflexive and transitive frames, then $\mathcal{F}^{S4} \models A$, and so $\vdash_{S4} A$.

Exercises 3.9

(1) KD is determined by the class of serial frames.

(2) $S5$ is determined by the class of equivalence relations (reflexive, transitive, and symmetric frames).

(3) $K4.3$ is determined by the class of transitive weakly-connected frames, and $S4.3$ by the class of reflexive, transitive, and weakly-connected frames.

(4) $S4.2$ is the name of the smallest normal logic containing $S4$ and the schema

$$\Diamond\Box A \to \Box\Diamond A.$$

Prove that $\mathcal{F}^{S4.2}$ is weakly-directed, and that $S4.2$ is determined by the class of reflexive, transitive, and weakly-directed frames.

(5) Axiomatise the logics determined by
  (i) the class of partially-functional frames;
  (ii) the class of functional frames;
  (iii) the class of weakly dense frames.

(6) (Harder). For all $n \geq 0$, define the formulae $\Box^n A$ inductively by

$$\Box^0 A = A$$
$$\Box^{n+1} A = \Box\Box^n A.$$

Thus $\Box^n A = \underbrace{\Box \ldots \Box}_{n \text{ times}} A$. Define the formula $\Diamond^n A$ similarly.

(i) Show that in any model $\mathcal{M}$,

$$\begin{array}{lll} \mathcal{M} \models_s \Box^n A & \text{iff} & sR^n t \text{ implies } \mathcal{M} \models_t A; \\ \mathcal{M} \models_s \Diamond^n A & \text{iff} & \exists t(sR^n t \ \& \ \mathcal{M} \models_t A). \end{array}$$

(ii) If Λ is a normal logic, show that

$$\vdash_\Lambda \Box^n A \wedge \Box^n B \leftrightarrow \Box^n(A \wedge B).$$

(iii) For any normal logic Λ, if $s, t \in S^\Lambda$, prove that

$$s(R^\Lambda)^n t \quad \text{iff} \quad \{A : \Box^n A \in s\} \subseteq t \quad \text{iff} \quad \{\Diamond^n A : A \in t\} \subseteq s.$$

(iv) For fixed k, l, m, n, let Λ contain the schema

$$\Diamond^k \Box^l A \to \Box^m \Diamond^n A.$$

Show that

$$s(R^\Lambda)^k t \ \& \ s(R^\Lambda)^m u \quad \text{implies} \quad \exists v(t(R^\Lambda)^l v \ \& \ u(R^\Lambda)^n v).$$

(v) Show how (iv) encompasses all the completeness theorems we have mentioned in the above theorems and exercises, except for $K4.3$ and $S4.3$.

$S5$: Logical Necessity and Introspective Knowledge

$S5$ is amongst the most well-known of modal logics, and is often regarded as the system which characterises the notion of *logical* necessity. It might be argued that a possible world, representing a different way the world could have been, ought to satisfy all the logical laws of the actual world, and so a context in which one of our logical laws was violated should not count as a possible world at all. From this point of view, a logically necessary truth is one which is true in all possible worlds whatsoever, suggesting the semantic analysis

$$\mathcal{M} \models_s \Box A \quad \text{iff} \quad \text{for all } t \in S, \ \mathcal{M} \models_t A.$$

But this is equivalent to confining our relational semantics to frames $\mathcal{F} = (S, R)$ in which R is *universal*, i.e. $R = S \times S$, so that any s has sRt for all $t \in S$.

Theorem 3.10. *S5 is determined by the class of universal frames.*

Proof. Soundness is left as an exercise. For completeness, suppose $\nvdash_{S5} A$. Then A is false at some point t in the canonical model $\mathcal{M}^{S5}$. But then by the Submodel Lemma 1.7, A is false at t in the the submodel of $\mathcal{M}^{S5}$ generated by t. This submodel is based on the set

$$\{u \in S^{S5} : t(R^{S5})^{*}u\},$$

where $(R^{S5})^{*}$ is the reflexive transitive closure of R^{S5}. Since R^{S5} is reflexive and transitive, this set is just

$$\{u : tR^{S5}u\},$$

the equivalence class of t under the equivalence relation R^{S5}. But an equivalence relation is universal on each of its equivalence classes.

The system $S5$ has been the focus of attention in work on the theory of computation relating to the representation of knowledge and information possessed by robotic systems and other "agents". Among the theorem-schemata of $S5$ are

$$\Box A \to \Box\Box A$$

$$\neg\Box A \to \Box\neg\Box A$$

(the latter being a variant of the schema 5). Reading $\Box B$ as "the agent knows B", the first of these says that if an agent knows something, then it knows that it knows it, while the second states that if it does not know something, then it knows that it does not know it. The principles of $S5$ are relevant to the study of an agent that possesses full introspection as to the content of its own knowledge. For further details of this application, cf. Parikh [1984] and Rosenchein [1985]. The paper of Rosenchein and Kaelbling [1986] presents a system in this context with modal connectives for time, necessity, and knowledge.

Connectedness

A frame is *connected* if it satisfies

$$\forall s \forall t(sRt \vee s = t \vee tRs).$$

This property is satisfied by $(S, <)$, where S is any of the number-sets ω, $\mathbb{Z}$, $\mathbb{Q}$, or $\mathbb{R}$, and the notion of connectedness will be of most interest to us in frames, such as these examples, that are also transitive. Any connected frame is *weakly* connected, but whereas the class of weakly-connected frames is characterised by the schema

$$L: \quad \Box(A \wedge \Box A \to B) \vee \Box(B \wedge \Box B \to A),$$

there is no similar schema that is valid in precisely the connected frames. To see this, take two connected frames $\mathcal{F}_1$ and $\mathcal{F}_2$ that have no elements in common, and form their union

$$\mathcal{F}_1 \uplus \mathcal{F}_2 = (S_1 \cup S_2, R_1 \cup R_2).$$

$\mathcal{F}_1 \uplus \mathcal{F}_2$ is called a *disjoint union*, since S_1 and S_2 are disjoint. It can be shown that any formula valid in both $\mathcal{F}_1$ and $\mathcal{F}_2$ will be valid in $\mathcal{F}_1 \uplus \mathcal{F}_2$ (cf. Ex. 3.11(3) below). But the latter is only weakly connected, not connected.

Nonetheless, by using the generated-submodel construction in the way that was done for $S5$ in Theorem 3.10 above, we will be able to produce the connected frames we need.

Exercises 3.11

(1) Let $\mathcal{F}$ be a transitive weakly connected frame. Prove that any generated subframe of $\mathcal{F}$ is connected. Then prove that $K4.3$ is determined by the class of transitive connected frames, and $S4.3$ by the class of reflexive, transitive, and connected frames.

(2) A frame is *directed* if it satisfies

$$\forall s \forall t \exists u (sRu \wedge tRu).$$

Prove that the logic $S4.2$ of Exercise 3.9(4) is determined by the class of reflexive, transitive, and directed frames.

(3) (*Disjoint Unions in General.*) Let $\{\mathcal{F}_i : i \in I\}$ be a collection of frames $\mathcal{F}_i = (S_i, R_i)$ that are pairwise disjoint, i.e. $S_i \cap S_j = \emptyset$ for all $i \neq j \in I$. Let

$$\biguplus_I \mathcal{F}_i = (\bigcup_I S_i, \bigcup_I R_i).$$

Show that

$$\biguplus_I \mathcal{F}_i \models A \quad \text{iff} \quad \text{for all } i \in I,\ \mathcal{F}_i \models A.$$

The result of this last exercise offers a deeper explanation of why $S5$ is determined by both the class of universal frames and the class of equivalence relations: any equivalence relation is the disjoint union of its equivalence classes, each of which is a universal frame.

4 | Filtrations and Decidability

To show that a logic Λ is complete with respect to a class $\mathcal{C}$ of structures, one may try to show that if $\nvdash_\Lambda A$ then there is a member of $\mathcal{C}$ that rejects A. Now we know that there will be some point in the canonical model $\mathcal{M}^\Lambda$ at which A is false, but in its capacity as a falsifying model for a *particular* non-theorem A, $\mathcal{M}^\Lambda$ provides a good deal of superfluous information. To begin with, to calculate the truth-value of A at points in $\mathcal{M}^\Lambda$, we need only know the truth-values in $\mathcal{M}^\Lambda$ of the members of the set $Sf(A)$ of subformulae of A, whereas $\mathcal{M}^\Lambda$ provides truth-values for all formulae whatsoever. Moreover, if Φ is infinite, then S^Λ will be infinite (in fact uncountable), and so a point of $\mathcal{M}^\Lambda$ will in general be indistinguishable from many other points as to how it treats the *finitely* many members of $Sf(A)$. Thus we many as well identify points that assign the same truth- values to all members of $Sf(A)$. The identification process allows us to collapse $\mathcal{M}^\Lambda$, and to form a new falsifying model for A, one that has room for variety in its definition. This process, known as *filtration*, gives a way of proving certain technical results (finite model property, decidability) about certain logics Λ. But more importantly, it gives a new way of constructing models that comes into its own in cases where $\mathcal{M}^\Lambda$ is not in the desired class $\mathcal{C}$ for a completeness theorem.

Filtrations

Fix a model $\mathcal{M} = (S, R, V)$ and a set $\Gamma \subseteq Fma(\Phi)$ that is closed under subformulae, i.e.

$$B \in \Gamma \quad \text{implies} \quad Sf(B) \subseteq \Gamma.$$

For each $s \in S$, define

$$\Gamma_s = \{B \in \Gamma : \mathcal{M} \models_s B\},$$

and put

$$s \sim_\Gamma t \quad \text{iff} \quad \Gamma_s = \Gamma_t,$$

so that

$$s \sim_\Gamma t \quad \text{iff} \quad \text{for all } B \in \Gamma,\ \mathcal{M} \models_s B \text{ iff } \mathcal{M} \models_t B.$$

Then $\sim_\Gamma$ is an equivalence relation on S. Let

$$|s| = \{t \in S : s \sim_\Gamma t\}$$

be the $\sim_\Gamma$-equivalence class of s, and define

$$S_\Gamma = \{|s| : s \in S\}$$

to be the set of all such equivalence classes.

Lemma 4.1. *If Γ is finite, then S_Γ is finite and has at most 2^n elements, where n is the number of elements of Γ.*

Proof. Since $|s| = |t|$ iff $s \sim_\Gamma t$ iff $\Gamma_s = \Gamma_t$, putting

$$f(|s|) = \Gamma_s$$

gives a well-defined and one-to-one mapping of S_Γ into the set of subsets of Γ. Hence S_Γ has no more elements than there are subsets of Γ. But if Γ has n elements, then it has 2^n subsets.

Exercise 4.2

S_Γ can be finite even if Γ is not. Define Γ to be *finitely based over $\mathcal{M}$* if there exists a finite set Δ of formulae such that

$$\forall B \in \Gamma\ \exists B_0 \in \Delta\, (\mathcal{M} \models B \leftrightarrow B_0).$$

Show that S_Γ is finite if Γ is finitely based over $\mathcal{M}$.

Now let $\Phi_\Gamma = \Phi \cap \Gamma$ be the set of atomic formulae that belong to Γ, and define

$$V_\Gamma : \Phi_\Gamma \to 2^{S_\Gamma}$$

by putting

$$|s| \in V_\Gamma(p) \quad \text{iff} \quad s \in V(p)$$

whenever $p \in \Phi_\Gamma$ (since then $p \in \Gamma$, V_Γ is well-defined).

We are going to consider Φ_Γ-models of the form $\mathcal{M}' = (S_\Gamma, R', V_\Gamma)$ with the property that the truth-values of members of Γ in $\mathcal{M}$ and in $\mathcal{M}'$ are left invariant by the correspondence $s \mapsto |s|$. Reflection on what is required to make this work leads to the following definition.

A binary relation R' on S_Γ is called a *Γ-filtration of R* if it satisfies

(F1) if sRt, then $|s|R'|t|$; and

(F2) if $|s|R'|t|$, then for all B,

if $\Box B \in \Gamma$ and $\mathcal{M} \models_s \Box B$, then $\mathcal{M} \models_t B$.

Any Φ_Γ-model $\mathcal{M}' = (S_\Gamma, R', V_\Gamma)$ in which R' satisfies F1 and F2 is called a *Γ-filtration of the model $\mathcal{M}$*.

Filtration Lemma 4.3. *If $B \in \Gamma$, then for any $s \in S$,*

$$\mathcal{M} \models_s B \quad \text{iff} \quad \mathcal{M}' \models_{|s|} B.$$

Proof. An important exercise for the reader. The case $B = p \in \Phi$ is given by the definition of V_Γ. The inductive case for the truth-functional connectives is straightforward, while the case for $\Box$ uses F1 and F2. Note that the closure of Γ under subformulae is needed in order to be able to apply the induction hypothesis.

Exercise 4.4

Let Γ^b be the Boolean closure of Γ, i.e. the closure of Γ under the propositional connectives. Show that the Filtration Lemma holds for all $B \in \Gamma^b$.

Examples of Filtrations

1. The *smallest* filtration.

$$|s|R^\sigma|t| \quad \text{iff} \quad \exists s' \in |s|\,\exists t' \in |t|(s'Rt').$$

2. The *largest* filtration.

$$|s|R^\lambda|t| \quad \text{iff} \quad \text{for all } B,\ \Box B \in \Gamma\ \&\ \mathcal{M} \models_s \Box B \text{ implies } \mathcal{M} \models_t B.$$

3. The *transitive* filtration.

$$|s|R^\tau|t| \quad \text{iff} \quad \text{for all } B,\ \Box B \in \Gamma\ \&\ \mathcal{M} \models_s \Box B \text{ implies } \mathcal{M} \models_t \Box B \wedge B.$$

Exercises 4.5

(1) R^σ and R^λ are always Γ-filtrations of R.

(2) If R' is any Γ-filtration of R, then
$$R^\sigma \subseteq R' \subseteq R^\lambda$$
(hence the names *smallest* and *largest*).

(3) R^τ is transitive and satisfies F2. If R is transitive, then R^τ is a Γ-filtration of R.

(4) Define a *symmetric* relation on S_Γ that is a Γ-filtration of R whenever R is symmetric.

(5) Show that the following properties are preserved in passing from R to any Γ-filtration of R: *reflexive, serial, connected, directed.*

Theorem 4.6. *K is determined by the class of all finite frames. Moreover, if a formula A has n subformulae, then $\vdash_K A$ if, and only if, A is valid in all frames having at most 2^n elements.*

Proof. Suppose $\nvdash_K A$. Then there is a point s in some model $\mathcal{M}$ at which A is false (e.g. $\mathcal{M} = \mathcal{M}^K$). Let $\Gamma = Sf(A)$. Then Γ is closed under subformulae, so we can construct Γ-filtrations $\mathcal{M}' = (S_\Gamma, R', V_\Gamma)$ of $\mathcal{M}$ as above. By the Filtration Lemma 4.3, A is false at $|s|$ in any such model, and hence not valid in the frame (S_Γ, R'). The desired bound on the size of S_Γ is given by Lemma 4.1.

Decidability

A logic Λ has the *finite frame property* if it is determined by its finite frames, i.e.,

if $\nvdash_\Lambda A$, then there is a finite frame $\mathcal{F}$ with $\mathcal{F} \models \Lambda$ and $\mathcal{F} \not\models A$.

Theorem 4.6 showed that the smallest normal logic K has the finite frame property, but it showed more: a computable bound was given on the size of the invalidating frame for a given non-theorem. This implies that the property of K-theoremhood is *decidable*, i.e. that there is an algorithm for determining, for each formula A, whether or not $\vdash_K A$. If A has n subformulae, we simply check to see whether or not A is valid in all frames of size at most 2^n. Since a finite set has finitely many binary relations (2^{m^2} relations on an m-element set), there are only finitely many frames of size at most 2^n. Moreover, to determine whether A is valid on a finite frame $\mathcal{F}$, we need only look at models $V : \Phi_A \to 2^S$ on $\mathcal{F}$, where $\Phi_A = \Phi \cap Sf(A)$. But there are only finitely many such models on $\mathcal{F}$. Thus the whole checking procedure for validity of A in frames of size at most 2^n can be completed in a finite amount of time.

To consider the case of logics other than K, we will say that Λ has the *strong* finite frame property if there is a computable function g such that

if $\nvdash_\Lambda A$, then there is a finite Λ-frame that invalidates A and has at most $g(n)$ elements, where n is the number of subformulae of A.

Now in adapting the above decidability argument to Λ, there is an extra feature. In addition to deciding whether or not a given finite frame $\mathcal{F}$ validates A, we also have to decide whether or not $\mathcal{F} \models \Lambda$. If Λ is *finitely axiomatisable*, meaning that

$$\Lambda = K\Sigma_1 \dots \Sigma_n$$

for some finite number of schemata Σ_j, then the property "$\mathcal{F} \models \Lambda$" is decidable: it suffices to determine whether each Σ_j is valid in $\mathcal{F}$. For all of the logics we have considered thus far, validity of Σ_j is equivalent to some first-order property of R, which can be algorithmically decided for finite $\mathcal{F}$. But in any case, validity of a schema, and hence of a finite number of schemata, on a finite frame, is always decidable. The point is that a schema, such as

$$\Box(A \land \Box A \to B) \lor \Box(B \land \Box B \to A),$$

has only finitely many "atomic components" $A, B, \ldots$, and there are only finitely many choices for the "truth-sets"

$$\{s : \mathcal{M} \models_s A\}, \quad \{s : \mathcal{M} \models_s B\}, \quad \ldots\ldots$$

of these components in all possible models $\mathcal{M}$ on $\mathcal{F}$. To put it another way: a schema is the set Σ_A of all substitution instances of some formula A, and validity of all members of Σ_A in frame $\mathcal{F}$ is equivalent (by 2.10(2)) to validity of A in $\mathcal{F}$, which, as noted on the previous page, is decidable when $\mathcal{F}$ is finite. Thus we have

Theorem 4.7. *Every finitely axiomatisable logic with the strong finite frame property is decidable.*

Exercises 4.8

(1) Prove that the logics KD, KT, $K4$, KB, $S4$, $S5$, $K4.3$, $S4.3$, $S4.2$ (Exercises 3.9(4), 3.11(2)), are all decidable.

(2) In fact any finitely axiomatisable logic with the finite frame property is decidable (i.e. the result holds without invoking the computable function g). Prove this as follows.

  (i) Show that a finitely axiomatisable logic Λ is effectively enumerable, i.e. there is an algorithm for enumerating the members of Λ (hint: cf. Exercise 2.8(1)).

  (ii) Show that if Λ has the finite frame property and is finitely axiomatisable, then the complement $Fma(\Phi) - \Lambda$ of Λ is effectively enumerable (hint: enumerate all the finite Λ-frames and systematically test formulae for validity in them).

  (iii) Use the fact that Λ is decidable iff both Λ and $Fma(\Phi) - \Lambda$ are effectively enumerable.

Finite Model Property

The topic of decidability could also be approached via the notion of the finite *model* property, which states that

if $\nvdash_\Lambda A$, then there is a finite Λ-model $\mathcal{M}$ with $\mathcal{M} \not\models A$.

It turns out that *for logics that are uniform*, this is equivalent to the finite frame property. The following exercises indicate how to prove this.

Exercises 4.9

A model $\mathcal{M}$ is *distinguished* if for any two distinct points s and t in $\mathcal{M}$ there is a formula A with $\mathcal{M} \models_s A$ and $\mathcal{M} \not\models_t A$.

(1) Show that any filtration is distinguished. Hence show that for any model $\mathcal{M}$, if $\Gamma = Fma(\Phi)$, then any Γ-filtration of $\mathcal{M}$ is a distinguished model that is "equivalent" to $\mathcal{M}$ in some suitable sense.

(2) If $\mathcal{M}$ is finite and distinguished, show that for each s in $\mathcal{M}$ there is a formula A_s such that for any t in $\mathcal{M}$,

$$\mathcal{M} \models_t A_s \quad \text{iff} \quad t = s.$$

(3) If $\mathcal{M}$ is finite and distinguished, show that for any subset X of $\mathcal{M}$ there exists a formula A_X such that for any t in $\mathcal{M}$,

$$\mathcal{M} \models_t A_X \quad \text{iff} \quad t \in X.$$

(4) Let $\mathcal{M}$ be a distinguished model on a finite frame $\mathcal{F}$. If $\mathcal{M}' = (\mathcal{F}, V')$ is any other model on $\mathcal{F}$, show that for all formulae A,

$$\mathcal{M} \models_s A' \quad \text{iff} \quad \mathcal{M}' \models_s A,$$

where A' is the result of uniformly replacing each atomic p in A by A_X, where $X = V'(p)$.
Deduce that for any *uniform* logic Λ,

$$\mathcal{M} \models \Lambda \quad \text{iff} \quad \mathcal{F} \models \Lambda.$$

(5) Complete the argument showing that for uniform logics, the finite model property implies the finite frame property .

Decidability Without the Finite Model Property

Although the finite frame property is sufficient to guarantee decidability for a finitely axiomatised logic, it is not necessary. The sharpest result in this direction would appear to be that of Cresswell [1984], which presents an example of a uniform normal logic that is finitely axiomatisable and decidable, but is *incomplete*, i.e. not determined by any class of frames at all. Such a logic cannot have the finite frame (or model) property.

The proof that Cresswell's logic is decidable uses a technique of translation into a decidable fragment of monadic second-order logic, and is beyond our scope.

5 | Multimodal Languages

Syntax

The whole theory presented so far adapts readily to languages with more than one modal connective. Given a set Φ of atomic formulae p, and a new collection of symbols $\{[\,i\,] : i \in I\}$, a set $Fma_I(\Phi)$ of formulae A is generated by the BNF definition

$$A ::= p \mid \bot \mid A_1 \to A_2 \mid [\,i\,]A,$$

so that we now have formulae $[\,i\,]A$ for each $A \in Fma_I(\Phi)$ and each $i \in I$. The connective $[\,i\,]$ is to be treated in the way we treated $\Box$ previously. The dual connective $<i>$ is defined as $\neg[\,i\,]\neg$, and corresponds to $\Diamond$.

Semantics

A *frame* for this new language is a structure

$$\mathcal{F} = (S, \{R_i : i \in I\}),$$

comprising a set S with a collection of binary relations $R_i \subseteq S \times S$, one for each $i \in I$. (Equivalently, we may think of $\mathcal{F}$ as a pair (S, R) with $R : I \to 2^{S \times S}$). A model $\mathcal{M} = (\mathcal{F}, V)$ on $\mathcal{F}$ is given by a function $V : \Phi \to 2^S$, just as before. The definition of the relation $\mathcal{M} \models_s A$ has the one new clause

$$\mathcal{M} \models_s [\,i\,]A \quad \text{iff} \quad \text{for all } t \in S,\ sR_it \text{ implies } \mathcal{M} \models_t A,$$

and the definitions of truth in a model ($\mathcal{M} \models A$), and validity in a frame ($\mathcal{F} \models A$), are unchanged.

Logics

The notion of tautology is defined as previously, taking all formulae of the form $[\,i\,]A$, along with members of Φ, in the definition of "quasi-atomic" formula.

A *logic* continues to be a subset Λ of $Fma_I(\Phi)$ that includes all tautologies and is closed under Detachment. The theory of *Λ-deducibility* and *Λ-maximal sets* then goes through without any changes.

A logic Λ is *normal* if it contains the schemata

$$K_i : [\,i\,](A \to B) \to ([\,i\,]A \to [\,i\,]B),$$

and satisfies the Necessitation rules

$$\vdash_\Lambda A \quad \text{implies} \quad \vdash_\Lambda [\,i\,]A,$$

for every $i \in I$. The smallest normal logic will be denoted K_I.

Canonical Model

For a normal logic Λ, the model

$$\mathcal{M}^\Lambda = (S^\Lambda, \{R_i^\Lambda : i \in I\}, V^\Lambda)$$

has

$$sR_i^\Lambda t \quad \text{iff} \quad \{B : [\,i\,]B \in s\} \subseteq t,$$

with the definitions of S^Λ and V^Λ remaining the same. The proof of the Truth Lemma for $\mathcal{M}^\Lambda$, i.e.,

$$\mathcal{M}^\Lambda \models_s A \quad \text{iff} \quad A \in s,$$

continues to work as previously: we simply treat each connective $[\,i\,]$ in the way we treated $\Box$ in §3. It follows that the logic K_I is determined by the class of all frames for the present language.

Filtrations

In defining a Γ-filtration $\mathcal{M}' = (S_\Gamma, \{R'_i : i \in I\}, V_\Gamma)$ of a model $\mathcal{M} = (S, \{R_i : i \in I\}, V)$, we stipulate, for each $i \in I$, that R'_i is a Γ-filtration of R_i iff

(F1$_i$) if sR_it, then $|s|R'_i|t|$; and

(F2$_i$) if $|s|R'_i|t|$, then for all B,
if $[\,i\,]B \in \Gamma$ and $\mathcal{M} \models_s [\,i\,]B$, then $\mathcal{M} \models_t B$.

Everything else, including the proof of the Filtration Lemma

$$\forall B \in \Gamma,\ \mathcal{M} \models_s B \quad \text{iff} \quad \mathcal{M}' \models_{|s|} B,$$

is as before. This yields a proof that the smallest normal logic K_I has the strong finite frame property and is decidable.

Generated Submodels

Given a model $\mathcal{M} = (S, \{R_i : i \in I\}, V)$, and an element $t \in S$, the *submodel* $\mathcal{M}^t = (S^t, \{R_i^t : i \in I\}, V^t)$ *generated by* t is defined as follows.

A subset X of S is *I-closed* if it satisfies:

if $u \in X$, then $v \in X$ whenever there is an $i \in I$ with uR_iv.

An intersection of I-closed sets is I-closed, so we can define S^t as the smallest I-closed subset of S that contains t. R_i^t and V^t are the restrictions of R_i and V to S^t.

Exercises 5.1

(1) Show that $\mathcal{M}^t \models_u A$ iff $\mathcal{M} \models_u A$.

(2) Show that for languages with a single modal connective (i.e. when I is a singleton), the present definition of $\mathcal{M}^t$ agrees with that given in §1.

(3) **p-Morphisms**. Formulate the appropriate notion of p-morphism for multimodal languages, and prove the analogues of the p-Morphism Lemmas 1.9 and 1.10.

6 | Temporal Logic

Consider a propositional language with two modal connectives, [F] and [P], meaning, respectively, *henceforth* (at all future times), and *hitherto* (at all past times). According to §5, a frame for this language has the form $(S, R_{\mathrm{F}}, R_{\mathrm{P}})$, with the modelling

$$\mathcal{M} \models_s [\mathrm{F}]A \quad \text{iff} \quad sR_{\mathrm{F}}t \text{ implies } \mathcal{M} \models_t A,$$

$$\mathcal{M} \models_s [\mathrm{P}]A \quad \text{iff} \quad sR_{\mathrm{P}}t \text{ implies } \mathcal{M} \models_t A.$$

We read $sR_{\mathrm{F}}t$ as "t is in the future of s" and $sR_{\mathrm{P}}t$ as "t is in the past of s". But the intended interpretation is that [F] and [P] express properties of the same time-ordering, so that t should be in the past of s precisely when s is in the future of t. Thus we want

$$sR_{\mathrm{P}}t \quad \text{iff} \quad tR_{\mathrm{F}}s$$

(or, equivalently, that the relations R_{P} and R_{F} are each the converse of the other).

Exercise 6.1

Let $\mathcal{F} = (S, R_{\mathrm{F}}, R_{\mathrm{P}})$.

(1) Show that

$$\mathcal{F} \models A \rightarrow [\mathrm{P}] <\mathrm{F}> A \quad \text{iff} \quad \forall s \forall t (sR_{\mathrm{P}}t \text{ implies } tR_{\mathrm{F}}s).$$

(2) Show that

$$\mathcal{F} \models A \rightarrow [\mathrm{F}] <\mathrm{P}> A \quad \text{iff} \quad \forall s \forall t (tR_{\mathrm{F}}s \text{ implies } sR_{\mathrm{P}}t).$$

(3) If a normal logic Λ contains the schema

$$A \rightarrow [\mathrm{P}] <\mathrm{F}> A,$$

then in $\mathcal{M}^{\Lambda}$,

$$sR^{\Lambda}_{\mathrm{P}}t \quad \text{implies} \quad tR^{\Lambda}_{\mathrm{F}}s.$$

(4) If a normal logic Λ contains the schema

$$A \rightarrow [\mathrm{F}] <\mathrm{P}> A$$

then

$$tR^{\Lambda}_{\mathrm{F}}s \quad \text{implies} \quad sR^{\Lambda}_{\mathrm{P}}t.$$

Temporal Logics

The preceding exercises indicate that any *temporal* logic should at least contain the two schemata that they discuss. In the frames for such a logic, R_{F} and R_{P} are interdefinable, so we may as well take one relation as primitive, and use frames $\mathcal{F} = (S, R)$, where $R \subseteq S \times S$, with the modelling

$$\mathcal{M} \models_s [\mathrm{F}]A \quad \text{iff} \quad sRt \text{ implies } \mathcal{M} \models_t A,$$

$$\mathcal{M} \models_s [\mathrm{P}]A \quad \text{iff} \quad tRs \text{ implies } \mathcal{M} \models_t A.$$

But it is natural also to require a temporal ordering to be *transitive*, so we will now define a *time-frame* to be any structure $\mathcal{F} = (S, R)$ with R a transitive relation on S, and with the modelling just given. A *temporal logic* is defined to be any normal logic in the language of $[\mathrm{F}]$ and $[\mathrm{P}]$ that contains the schemata

$$\begin{array}{ll} C_{\mathrm{P}}: & A \to [\mathrm{P}] <\mathrm{F}> A \\ C_{\mathrm{F}}: & A \to [\mathrm{F}] <\mathrm{P}> A \\ 4_{\mathrm{P}}: & [\mathrm{P}]A \to [\mathrm{P}][\mathrm{P}]A \\ 4_{\mathrm{F}}: & [\mathrm{F}]A \to [\mathrm{F}][\mathrm{F}]A \end{array}$$

Mirror Images

Notice that these schemata come in pairs, related by interchanging past and future connectives. Members of such pairs are called "mirror images" of each other.

The smallest temporal logic, which is

$$K_{\{\mathrm{P},\mathrm{F}\}} C_{\mathrm{P}} C_{\mathrm{F}} 4_{\mathrm{P}} 4_{\mathrm{F}}$$

in the present notation, is commonly known as K_t in the literature.

Exercises 6.2

(1) Prove that K_t is determined by the class of all time-frames.

(2) Show that only one of 4_{P} and 4_{F} is needed in the definition of K_t: each is derivable from the remaining axioms.

Always

One way to view temporal logic is as a more powerful language for expressing properties of frames of the form (S, R). To this end it is useful to introduce the connective $\Box$ by definitional abbreviation, writing $\Box A$ for the formula

$$[\mathrm{P}]A \wedge A \wedge [\mathrm{F}]A.$$

$\Box A$ may be read "*always* A", i.e. at all times, past, present, and future. The dual formula $\Diamond A = \neg\Box\neg A$ is tautologically equivalent to

$$<\mathrm{P}> A \vee A \vee <\mathrm{F}> A,$$

meaning "at some time (past, present, or future), A".

Exercises 6.3

Let $\mathcal{F}$ be any frame.

(1) Show that

$$\mathcal{F} \models <\mathrm{P}> A \to [\mathrm{F}] <\mathrm{P}> A$$

iff R is transitive. What is the mirror image of this result?

(2) Show that

$$\mathcal{F} \models \Box A \to [\mathrm{P}][\mathrm{F}]A$$

iff R is *weakly future-connected*, i.e.

$$sRt \wedge sRu \to (tRu \vee t = u \vee uRt).$$

(3) If Λ contains C_{P}, C_{F}, and the schema

$$\Box A \to [\mathrm{P}][\mathrm{F}]A,$$

show that $\mathcal{F}^{\Lambda}$ is weakly future-connected.

(4) Work out the mirror images of Exercises 2 and 3.

(5) Explain why $\Box$ behaves like an $S5$ modality in a *connected* frame.

Strict and Total Orderings

- A *strict ordering* is a time-frame whose transitive relation is *irreflexive*, and hence has the stronger property of *asymmetry* (cf. Exercise 1.15).
- A *total ordering* is a time-frame whose transitive relation is *connected* and *antisymmetric*, like the numerical orderings $<$ and $\leq$ on $\mathbb{R}$.
- A *strict total ordering* is therefore an irreflexive total ordering, or, more simply, a relation that is transitive, connected, and irreflexive.

We tend to use the symbol $<$ for the relation of a strict ordering. An *immediate successor* of an element x is an element y with $x < y$ and such that there is no z with $x < z < y$. A *cut* in a structure $(S, <)$ is a partition of S into a pair (X, Y) of non-empty disjoint subsets with $x < y$ whenever $x \in X$ and $y \in Y$. A strict total ordering is *continuous* if for any such cut there is a z with $x \leq z \leq y$ for all $x \in X$ and $y \in Y$ (where $x \leq z$ iff $x < z$ or $x = z$). Intuitively, this means that there are no "gaps" in the ordering.

Exercises 6.4

Let $\mathcal{F}$ be a strict total ordering.

(1) Show that

$$\mathcal{F} \models A \wedge [\mathrm{P}]A \to <\mathrm{F}> [\mathrm{P}]A$$

iff every element of $\mathcal{F}$ has an immediate successor.

(2) Show that

$$\mathcal{F} \models A \wedge [\mathrm{P}]A \rightarrow [\mathrm{F}]\bot \vee \langle\mathrm{F}\rangle [\mathrm{P}]A$$

iff every element except the last one (if it exists) has an immediate successor. (An element x is *last* if there is no y with $x < y$.)

(3) Work out the mirror images of Exercises 1 and 2.

(4) Show that

$$\mathcal{F} \models \Box([\mathrm{P}]A \rightarrow \langle\mathrm{F}\rangle [\mathrm{P}]A) \rightarrow ([\mathrm{P}]A \rightarrow [\mathrm{F}]A)$$

iff $\mathcal{F}$ is continuous.

This last exercise demonstrates that temporal logic is "more expressive" that the language we began with in §1. The real-number frame $(\mathbb{R}, <)$ is continuous, while the rational-number frame $(\mathbb{Q}, <)$ is not. But when these are used as frames for the language of a single modal connective, the same formulae are valid in each: consult Exercises 8.8 in the next Part to see how this is proved. We will also see at the end of §8 how to use the schema of Exercise (4) above in a completeness proof for the temporal logic of $(\mathbb{R}, <)$.

Generated Time Models

According to the definition given in §5, if $\mathcal{M} = (S, R, V)$ is a model on a time-frame, then the submodel $\mathcal{M}^t = (S^t, R^t, V^t)$ generated by an element $t \in S$ has S^t as the smallest subset X of S that contains t and is closed under R_{P} and R_{F}, which means that

if $u \in X$, then $v \in X$ whenever uRv or vRu.

Exercises 6.5

(1) Let $\bar{R} = R \cup R^{-1}$, where $R^{-1} = \{(v, u) : uRv\}$. Show that

$$S^t = \{u \in S : t(\bar{R})^* u\}.$$

(2) Suppose that R is weakly future-connected and weakly past-connected (cf. Ex. 6.3(2)). Show

$$S^t = \{u : tRu \text{ or } t = u \text{ or } uRt\}.$$

(3) Prove that if time-frame $\mathcal{F}$ is weakly future-connected and weakly past-connected, then the generated time-frame $\mathcal{F}^t = (S^t, R^t)$ is *connected*.

Temporal p-Morphisms

For temporal logic, a p-morphism $f : \mathcal{M}_1 \to \mathcal{M}_2$ must satisfy the conditions

$$\begin{array}{lll} sR_1t & \text{implies} & f(s)R_2f(t), \\ f(s)R_2u & \text{implies} & \exists t(sR_1t \ \& \ f(t) = u), \\ uR_2f(s) & \text{implies} & \exists t(tR_1s \ \& \ f(t) = u), \end{array}$$

in order for the p-Morphism Lemma

$$\mathcal{M}_1 \models_s A \quad \text{iff} \quad \mathcal{M}_2 \models_{f(s)} A$$

to hold for all formulae A in the language of [P] and [F].

Temporal Filtrations

In defining Γ-filtrations of models $\mathcal{M} = (S, R, V)$ on time-frames, we want to preserve both the transitivity of R and the fact that R is R_{F} and R^{-1} is R_{P}. A suitable relation for this purpose is $R^{\tau} \subseteq S_\Gamma \times S_\Gamma$, where

$$\begin{array}{lll} |s|R^{\tau}|t| & \text{iff} & [\mathrm{F}]B \in \Gamma \text{ and } \mathcal{M} \models_s [\mathrm{F}]B \text{ implies } \mathcal{M} \models_t [\mathrm{F}]B \wedge B, \\ & & \text{and} \\ & & [\mathrm{P}]B \in \Gamma \text{ and } \mathcal{M} \models_t [\mathrm{P}]B \text{ implies } \mathcal{M} \models_s [\mathrm{P}]B \wedge B. \end{array}$$

The model $\mathcal{M}^{\tau} = (S_\Gamma, R^{\tau}, V_\Gamma)$ is then transitive, and the Filtration Lemma

$$\mathcal{M} \models_s B \quad \text{iff} \quad \mathcal{M}^{\tau} \models_{|s|} B$$

holds for all $B \in \Gamma$.

Exercises 6.6

(1) Verify this last claim.

(2) Prove that the smallest temporal logic K_t is determined by the class of finite time-frames and is decidable.

(3) Axiomatise the logic determined by the class of *connected* time-frames, proving that it has the strong finite frame property and is decidable.

Diodorean Modality

The most common practice in temporal logic is to regard time as an irreflexive ordering, so that "henceforth", meaning "at all future times", does not refer to the present moment. On the other hand, the Greek philosopher Diodorus proposed that the necessary be identified with that which is now and will always be the case. This suggests a temporal interpretation of $\Box$ that is naturally formalised by using reflexive orderings. The same interpretation is adopted in the logic of concurrent programs to be discussed in §9.

The Diodorean analysis leads to the study of systems containing $S4$, and containing $S4.3$ in the case of total orderings. When time is regarded as an endless discrete total ordering, the resulting logic is a system known as $S4.3Dum$, which will be investigated in §8.

Minkowski Spacetime

The Diodorean logic of four-dimensional special-relativistic spacetime has been shown to be the system $S4.2$ of Exercise 3.9(4) (Goldblatt [1980]). To explain this further, we first describe the structure of n-dimensional spacetime.

If $x = \langle x_1, \ldots, x_n \rangle$ is an n-tuple of real numbers, let

$$\mu(x) = x_1^2 + \cdots + x_{n-1}^2 - x_n^2.$$

Then *n-dimensional spacetime*, for $n \geq 2$, is the frame

$$\mathbb{T}^n = (\mathbb{R}^n, \leq),$$

where $\mathbb{R}^n$ is the set of all real n-tuples, and for x and y in $\mathbb{R}^n$ we have

$$\begin{aligned} x \leq y \quad &\text{iff} \quad \mu(y - x) \leq 0 \;\&\; x_n \leq y_n \\ &\text{iff} \quad \sum_{i=1}^{n-1}(y_i - x_i)^2 \leq (y_n - x_n)^2 \;\&\; x_n \leq y_n. \end{aligned}$$

The Minkowski spacetime of special relativity theory is $\mathbb{T}^4$, in which a typical point represents a spatial location $\langle x_1, x_2, x_3 \rangle$ at time x_4. The intended interpretation of the relation $x \leq y$ is that a signal can be sent from x to y at a speed at most that at which light travels, so that y is in the "causal future" of x.

The frame $\mathbb{T}^2$ is depicted in the following diagram, showing the "future cone" $\{z : x \leq z\}$ for a typical point x.

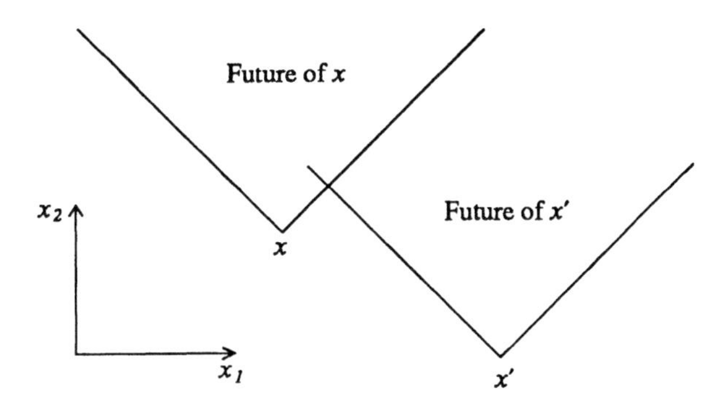

Observe that the future cones of any two points must overlap, so that the frame is *directed* and validates the $S4.2$ axiom schema

$$\Diamond\Box A \to \Box\Diamond A.$$

The work of Goldblatt [1980] involves showing that each of the frames $\mathbb{T}^n$ has $S4.2$ as its Diodorean modal logic. Noting that there is a natural p-morphism from $\mathbb{T}^{n+1}$ onto $\mathbb{T}^n$ (viz. delete the "first coordinate"), the heart of the proof is then a demonstration that there is a p-morphism from $\mathbb{T}^2$ onto any finite generated frame that is reflexive, transitive, and directed. The class of such finite frames determines $S4.2$ (Exercises 3.11(2), 4.5(5)).

Other interesting (strict) temporal orderings can be considered on spacetime, such as

$$x \prec y \quad \text{iff} \quad \mu(y - x) < 0 \;\&\; x_n < y_n,$$

and

$$x\alpha y \quad \text{iff} \quad x \neq y \;\&\; x \leq y.$$

Here $x \prec y$ holds when a signal can be sent from x to y at *slower* than light-speed, while α is the irreflexive version of $\leq$.

These orderings can be distinguished in terms of the validity of modal formulae. There may be two propositions A and B that are true in the future just at two points that can only be reached by travelling in different directions at the speed of light. Then $\Diamond A \wedge \Diamond B$ will be true now, but never again in the future. Thus the schema

$$\Diamond A \wedge \Diamond B \rightarrow \Diamond(\Diamond A \wedge \Diamond B)$$

is falsifiable under the ordering α. It is however valid under $\prec$, since a slower-than-light journey can always be speeded up, so we could wait some time and then travel at a greater speed to A and to B. This example is discussed further in Goldblatt [1980], where it is also shown that under α it is possible to distinguish the dimension of spacetime, e.g. there is a modal formula that is valid in $(\mathbb{R}^2, \alpha)$ but not in $(\mathbb{R}^3, \alpha)$.

The only known axiomatisation result for the temporal logic of spacetime is the one we have been discussing concerning $S4.2$. There are several open problems here that the reader may be interested in thinking about:

- axiomatise the full temporal logic, in the language of [F] and [P], of the frames $(\mathbb{R}^n, \leq)$;
- axiomatise the full temporal logic, and the Diodorean modal logic, of the frames $(\mathbb{R}^n, \prec)$ and $(\mathbb{R}^n, \alpha)$;
- analyse the case of *discrete* spacetime, in which $\mathbb{R}$ is replaced by $\mathbb{Z}$.

Since and Until

Consider a binary connective $\mathcal{U}$ with the semantics

$$\begin{aligned} \mathcal{M} \models_s A\mathcal{U}B \quad \text{iff} \quad & \text{there exists } t \text{ with } sRt \text{ and } \mathcal{M} \models_t B, \\ & \text{and } \mathcal{M} \models_u A \text{ for all } u \text{ such that } sRu \text{ and } uRt. \end{aligned}$$

The formula $A \mathcal{U} B$ is read "*A until B*", meaning that there is a future point at which B is true, with A true at all points between now and then.

Notice that

$$\mathcal{M} \models_s \top \mathcal{U} B \quad \text{iff} \quad \text{there exists } t \text{ such that } sRt \text{ and } \mathcal{M} \models_t B,$$

so that the formula $\top \mathcal{U} B$ is equivalent to $\langle \mathrm{F} \rangle B$. Hence $[\mathrm{F}]B$ is equivalent to $\neg(\top \mathcal{U} \neg B)$.

The formula $\bot \mathcal{U} B$ expresses that B will be true at a future point, with nothing in between, i.e. B is true at an immediate successor. Hence this formula is read "*next B*", and is a natural construct to consider on *discrete* orderings, like $(\mathbb{Z}, <)$ and $(\omega, <)$.

Exercise 6.7

Give a semantics for the notion "*A since B*", and use the notion to define $\langle \mathrm{P} \rangle B$ and a formula expressing "B was true at the previous moment".

The connectives *since* and *until* have been shown (Kamp [1968]) to form a *complete* set of connectives for continuous orderings. In a way that can be made precise, they suffice to define all possible propositional connectives that express temporal properties of such orderings. The connective *until* has been used extensively in the temporal logic of concurrent programs, and will be studied in that context in §9.

7 | Some Topics In Metatheory

We now take up some advanced topics: first-order definability, canonicity, incompleteness, and undecidability. (The material of this section is not needed in later sections.)

First-Order Definability

In §1 a number of examples were given of modal schemata whose frames were characterised by first-order conditions on a binary relation R. All of these, and many others, can be subsumed under a general class of schemata devised by Lemmon and Scott (Lemmon [1977]).

A formula φ is *positive* if it can be constructed using no connectives other than $\wedge$, $\vee$, $\Box$ and $\Diamond$. Thus a BNF definition of the class of positive formulae is

$$\varphi ::= p \mid \varphi_1 \wedge \varphi_2 \mid \varphi_1 \vee \varphi_2 \mid \Box\varphi \mid \Diamond\varphi.$$

We write $\varphi(p_1, \ldots, p_k)$ to indicate that the atomic formulae occurring in φ are among the list $p_1, \ldots, p_k$. $\varphi(A_1, \ldots, A_k)$ is then the formula obtained by uniformly substituting, for each $1 \leq i \leq k$, the formula A_i for p_i in φ.

Exercise 7.1

Let $\varphi(p_1, \ldots, p_k)$ be positive. If Λ is any normal logic, and $\vdash_\Lambda A_i \to B_i$ for $1 \leq i \leq k$, then

$$\vdash_\Lambda \varphi(A_1, \ldots, A_k) \to \varphi(B_1, \ldots, B_k).$$

Recall from Exercise 3.9(6) the notations

$$\Box^n A = \underbrace{\Box \ldots \Box}_{n \text{ times}} A,$$

$$\Diamond^n A = \underbrace{\Diamond \ldots \Diamond}_{n \text{ times}} A.$$

Then for each positive formula $\varphi(p_1, \ldots, p_k)$, and pair $\mathbf{m} = \langle m_1, \ldots, m_k \rangle$ and $\mathbf{n} = \langle n_1, \ldots, n_k \rangle$ of k-tuples of natural numbers, there is an associated *Lemmon-Scott schema*

$$\varphi^{\mathbf{m}}_{\mathbf{n}}: \quad \Diamond^{m_1}\Box^{n_1}A_1 \wedge \cdots \wedge \Diamond^{m_k}\Box^{n_k}A_k \to \varphi(A_1, \ldots, A_k).$$

Corresponding to this schema is a first-order condition $R\varphi^{\mathbf{m}}_{\mathbf{n}}$ on binary relations R. First, for a given frame $\mathcal{F} = (S, R)$ and a k-tuple $\mathbf{t} = \langle t_1, \ldots, t_k \rangle$ of elements of S, a condition $R_\varphi(s, \mathbf{t}, \mathbf{n})$ on $\mathcal{F}$, with "free variable" s, is defined by induction on the formation of the positive formula φ, as follows.

$$
\begin{array}{lll}
R_{p_i}(s,\mathbf{t},\mathbf{n}) & \text{is} & t_i R^{n_i} s \qquad (1 \le i \le k) \\
R_{\varphi_1 \wedge \varphi_2}(s,\mathbf{t},\mathbf{n}) & \text{is} & R_{\varphi_1}(s,\mathbf{t},\mathbf{n}) \wedge R_{\varphi_2}(s,\mathbf{t},\mathbf{n}) \\
R_{\varphi_1 \vee \varphi_2}(s,\mathbf{t},\mathbf{n}) & \text{is} & R_{\varphi_1}(s,\mathbf{t},\mathbf{n}) \vee R_{\varphi_2}(s,\mathbf{t},\mathbf{n}) \\
R_{\Box\varphi}(s,\mathbf{t},\mathbf{n}) & \text{is} & \forall u(sRu \to R_\varphi(u,\mathbf{t},\mathbf{n})) \\
R_{\Diamond\varphi}(s,\mathbf{t},\mathbf{n}) & \text{is} & \exists u(sRu \wedge R_\varphi(u,\mathbf{t},\mathbf{n})).
\end{array}
$$

Then $R\varphi^{\mathbf{m}}_{\mathbf{n}}$ is the first-order condition

$$\forall s \forall t_1 \cdots \forall t_k (sR^{m_1} t_1 \wedge \cdots \wedge sR^{m_k} t_k \to R_\varphi(s,\mathbf{t},\mathbf{n})).$$

Exercises 7.2

(1) In any model $\mathcal{M}$, if $R_\varphi(s,\mathbf{t},\mathbf{n})$ and $\mathcal{M} \models_{t_i} \Box^{n_i} A_i$ for $1 \le i \le k$, then $\mathcal{M} \models_s \varphi(A_1, \ldots, A_k)$.
(2) If $\mathcal{F}$ satisfies $R\varphi^{\mathbf{m}}_{\mathbf{n}}$, then $\mathcal{F} \models \varphi^{\mathbf{m}}_{\mathbf{n}}$.
(3) In any model $\mathcal{M}$, if $\mathcal{M} \models_s \varphi(p_1, \ldots, p_k)$, and $V(p_i) = \{u : t_i R^{n_i} u\}$ for $1 \le i \le k$, then $R_\varphi(s,\mathbf{t},\mathbf{n})$.
(4) If $\mathcal{F} \models \varphi^{\mathbf{m}}_{\mathbf{n}}$, then $\mathcal{F}$ satisfies $R\varphi^{\mathbf{m}}_{\mathbf{n}}$.

These exercises show that the frames validating $\varphi^{\mathbf{m}}_{\mathbf{n}}$ are precisely those satisfying $R\varphi^{\mathbf{m}}_{\mathbf{n}}$, and hence in particular that the logic $K\varphi^{\mathbf{m}}_{\mathbf{n}}$ is sound with respect to these frames. Completeness can be shown by the canonical model method, with the key result being

Lemma 7.3. *If $\varphi(p_1, \ldots, p_k)$ is positive, then the canonical frame for any normal logic Λ satisfies*

$$R^\Lambda_\varphi(s,\mathbf{t},\mathbf{n}) \quad \text{iff} \quad \{\varphi(A_1, \ldots, A_k) : \Box^{n_1} A_1 \in t_1 \;\&\cdots\&\; \Box^{n_k} A_k \in t_k\} \subseteq s.$$

Proof.
By induction on the formation of φ. We give the proof for $k = 1$, and drop the subscripts. The case $\varphi = p$ amounts to the claim that

$$tR^n s \quad \text{iff} \quad \{A : \Box^n A \in t\} \subseteq s$$

which was given as Exercise 3.9(6)(iii).

The most complex part of the proof concerns the inductive case of $\Diamond\varphi$, under the hypothesis that the Lemma holds for φ. Assuming that

$$\{\Diamond\varphi(A) : \Box^n A \in t\} \subseteq s, \tag{†}$$

we have to show that $R_{\Diamond\varphi}(s,t,n)$, i.e. that there exists a $u \in S^{\Lambda}$ with $sR^{\Lambda}u$ and $R_{\varphi}(u,t,n)$. But then it suffices to show that the set

$$u_0 = \{A : \Box A \in s\} \cup \{\varphi(B) : \Box^n B \in t\}$$

is Λ-consistent. For, if u is an Λ-maximal extension of u_0, then the definition of R^{Λ} and the induction hypothesis on φ ensure that u has the desired properties.

Now if u_0 is not Λ-consistent, then since $\{A : \Box A \in s\}$ is closed under finite conjunctions, it follows that there are formulae $A, B_1, \ldots, B_l$ such that $\Box A \in s$, $\Box^n B_i \in t$ for $1 \leq i \leq l$, and

$$\vdash_{\Lambda} A \to \neg(\varphi(B_1) \wedge \cdots \wedge \varphi(B_l)).$$

Hence

$$\vdash_{\Lambda} \Box A \to \Box\neg(\varphi(B_1) \wedge \cdots \wedge \varphi(B_l)).$$

Since $\Box A \in s$, it follows that

$$\neg\Diamond(\varphi(B_1) \wedge \cdots \wedge \varphi(B_l)) \in s. \qquad (\ddagger)$$

Now let $B = B_1 \wedge \cdots \wedge B_n$. Then it may be shown that $\Box^n B \in t$ (cf. Exercise 3.9(6)(ii)), and so by (†), $\Diamond\varphi(B) \in s$. But $\vdash_{\Lambda} \varphi(B) \to \varphi(B_i)$ for $1 \leq i \leq l$, by Exercise 7.1, so

$$\vdash_{\Lambda} \varphi(B) \to \varphi(B_1) \wedge \cdots \wedge \varphi(B_l),$$

whence

$$\vdash_{\Lambda} \Diamond\varphi(B) \to \Diamond(\varphi(B_1) \wedge \cdots \wedge \varphi(B_l)),$$

and thus

$$\Diamond(\varphi(B_1) \wedge \cdots \wedge \varphi(B_l)) \in s,$$

which is impossible, given (‡) and the Λ-consistency of s. Therefore, u_0 must be Λ-consistent as desired.

The proof that $R_{\Diamond\varphi}(s,t,n)$ implies $\{\Diamond\varphi(A) : \Box^n A \in t\} \subseteq s$ is straightforward, as are the inductive cases for $\varphi_1 \wedge \varphi_2$ and $\Box\varphi$. The case of $\varphi_1 \vee \varphi_2$ makes a further use of Exercise 7.1, and is also left to the reader (cf. Goldblatt [1975i]).

Exercises 7.4.

(1) Complete the proof of Lemma 7.3.

(2) Let Λ be a normal logic that contains the schema $\varphi^{\mathbf{m}}_{\mathbf{n}}$. Prove that $\mathcal{F}^{\Lambda}$ satisfies the first-order condition $R\varphi^{\mathbf{m}}_{\mathbf{n}}$.

(3) If Λ is the smallest normal logic containing a collection $\{(\varphi_i)^{\mathbf{m}_i}_{\mathbf{n}_i} : i \in I\}$ of Lemmon-Scott schemata, show that Λ is determined by the class of those frames that satisfy all the conditions $\{R(\varphi_i)^{\mathbf{m}_i}_{\mathbf{n}_i} : i \in I\}$.

Sahlqvist's Schemata

The form of the schema $\varphi^{\mathbf{m}}_{\mathbf{n}}$ was generalised by Sahlqvist [1975], to consider formulae of the type

$$\Box^n(\varphi \to \psi),$$

where $n \geq 0$, ψ is positive, and φ is constructed from atomic formulae and/or their negations using at most $\wedge$, $\vee$, $\Box$ and $\Diamond$, in such a way that no occurrence of $\wedge$, $\vee$ or $\Diamond$ is inside the scope of a $\Box$.

Sahlqvist showed that the frames validating any such formula are characterised by a first-order condition on R, and that this condition is satisfied by the canonical frame of the normal logic axiomatised by the schema corresponding to the formula.

A recent discussion and proof of this result may be found in Sambin and Vaccaro [1989].

Canonicity

A normal logic Λ is *canonical* if it is validated by its canonical frame, i.e. if $\mathcal{F}^\Lambda \models \Lambda$. The most accessible example of failure of canonicity is the logic KW, where W is the schema

$$\Box(\Box A \to A) \to \Box A$$

first mentioned in §1. Completeness for KW will be considered in §8 (cf. Exercises 8.7) where it is indicated that the logic has the finite frame property, so is determined by its (finite) frames and is decidable. The failure of canonicity is based on the following observation.

Exercise 7.5

Let $\mathcal{M} = (S, R, V)$ be a model containing a point s such that sRs. If $V(p) = S - \{s\}$, show that $\Box(\Box p \to p) \to \Box p$ is false at s in $\mathcal{M}$.

This Exercise shows that W is not valid on any frame possessing a *reflexive* point, and so to show that KW is not canonical it suffices to exhibit such an $s \in S^{KW}$ with $sR^{KW}s$. For this purpose, consider the set

$$s_0 = \{\neg\Box A : \nvdash_{KW} A\}.$$

If s_0 is KW-consistent, then any KW-maximal extension s of s_0 will solve the problem. For, if $A \notin s$, then $\nvdash_{KW} A$, and so as $s_0 \subseteq s$ and s is consistent, $\Box A \notin s$, showing that $sR^{KW}s$.

To prove that s_0 is KW-consistent, take formulae $A_1, \ldots, A_n$ such that $\nvdash_{KW} A_i$ for $1 \leq i \leq n$. We need to show that

$$\nvdash_{KW} \neg(\neg\Box A_1 \wedge \cdots \wedge \neg\Box A_n). \tag{i}$$

Now for each i there is some $s_i \in S^{KW}$ with $A_i \notin s_i$. Hence if $\mathcal{M}^i = (S^i, R^i, V^i)$ is the submodel of $\mathcal{M}^{KW}$ generated by s_i, then

$$\mathcal{M}^i \not\models_{s_i} A_i. \tag{ii}$$

We now construct a new model $\mathcal{M} = (S, R, V)$ by forming the union of all the models $\mathcal{M}^i$ and adjoining an additional element ∞ that is not in any of the S^i, but is R-related to all members of all the S^i. Formally, put

$$\begin{aligned} S &= S^1 \cup \cdots \cup S^n \cup \{\infty\} \\ R &= R^1 \cup \cdots \cup R^n \cup \{\langle\infty, s\rangle : s \in S \ \&\ \infty \neq s\} \\ V(p) &= V^1(p) \cup \cdots \cup V^n(p) \end{aligned}$$

Exercises 7.6

(1) If $s \in S^i$, then for any formula B,

$$\mathcal{M} \models_s B \quad \text{iff} \quad \mathcal{M}^i \models_s B.$$

(2) $\mathcal{M} \models W$
Note: in view of Exercise (1), the heart of the matter is to show that any instance of W is true in $\mathcal{M}$ at ∞.

In view of 7.6(2), the normal logic

$$\Lambda_{\mathcal{M}} = \{B : \mathcal{M} \models B\}$$

contains KW. But in view of (ii), 7.6(1) and the construction of $\mathcal{M}$,

$$\mathcal{M} \models_\infty \neg\Box A_1 \wedge \cdots \wedge \neg\Box A_n.$$

and so

$$\neg(\neg\Box A_1 \wedge \cdots \wedge \neg\Box A_n) \notin \Lambda_{\mathcal{M}}.$$

It follows that (i) must be true, completing the proof that s_0 is KW-consistent, and hence that KW is not canonical.

Canonicity and First-Order Definability

A class $\mathcal{C}$ of frames is *elementary* if there is a set of first-order conditions such that $\mathcal{C}$ comprises precisely those frames satisfying these conditions. A logic Λ is *first-order determined* if there is an elementary class $\mathcal{C}$ of frames such that Λ is the logic $\Lambda_{\mathcal{C}}$ determined by $\mathcal{C}$.

In many cases a logic can be shown to be first-order determined by showing that its canonical frame satisfies the first-order conditions in question (for instance, this applies to all the Sahlqvist schemata). In such cases the logic is also of course shown to be canonical. The universal relevance of this method is demonstrated by the following remarkable result of van Benthem [1980]:

> *if $\mathcal{C}$ is an elementary class of frames, then the logic $\Lambda_{\mathcal{C}}$ is canonical.*

van Benthem's proof is model-theoretic, using a compactness argument, and builds on earlier work of Fine [1975], which showed that a logic Λ is canonical if the class $\{\mathcal{F} : \mathcal{F} \models \Lambda\}$ of all its frames is closed under elementary equivalence. An alternative "structural" approach to van Benthem's result is given in Goldblatt [1990].

It seems plausible to conjecture that the converse result is true, i.e. that

> *if a logic Λ is canonical, then there is an elementary class $\mathcal{C}$ such that $\Lambda = \Lambda_{\mathcal{C}}$.*

A possible candidate for $\mathcal{C}$ here is the class of all frames that satisfy the same first-order conditions as $\mathcal{F}^{\Lambda}$. At present the conjecture is open.

The McKinsey Axiom

This is the schema

$$M: \qquad \Box\Diamond A \rightarrow \Diamond\Box A$$

mentioned in §1, where it was pointed out that the class of all frames for M is not elementary. It was shown further in Goldblatt [1976] that the logic KM is not determined by *any* elementary class of frames.

It would appear that the schema M is the simplest example not (equivalent to one) meeting the definition of Sahlqvist's schemata. Until recently the question of the canonicity of KM was unresolved, leaving open the possibility that it could be a counter-example to the above conjecture. However, that possibility has now been removed: a proof is given in Goldblatt [1991] that there is a model on $\mathcal{F}^{KM}$ that falsifies an instance of M.

In other respects KM is better behaved. Fine [1975i] shows that it has the finite frame property and is decidable.

Failure of the Finite Frame Property

If a logic has the finite frame property, then it complete with respect to the class of its validating frames. The converse of this is false: we now exhibit a logic that is first-order determined, but lacks the finite model property, and hence the finite frame property. The example is an adaptation by Hughes and Cresswell [1984] of a fundamental construction introduced by Makinson [1969].

Consider the Lemmon-Scott schema

$$Mk: \qquad \Box A_1 \wedge A_2 \rightarrow \Diamond(\Box^2 A_1 \wedge \Diamond A_2).$$

Exercise 7.7

Verify that the first-order condition corresponding to Mk is

$$\forall s \exists t(sRt \wedge tRs \wedge \forall u(tR^2u \rightarrow sRu)).$$

Now let Λ_* be the logic $KTMk$. Then from our analysis of Lemmon-Scott schemata we know that Λ_* is determined by the class of all reflexive frames that satisfy the condition of 7.7. To show that Λ_* lacks the finite model property we prove two things:

(1) If $\mathcal{M}$ is a *finite* Λ_*-model, then the schema 4 is true in $\mathcal{M}$, i.e. for any formula A, $\mathcal{M} \models \Box A \rightarrow \Box^2 A$.

(2) For some A, $\not\vdash_{\Lambda_*} \Box A \rightarrow \Box^2 A$.

Proof of (1). Let $\mathcal{M}$ be a Λ_*-model that rejects 4. Then we show that $\mathcal{M}$ must be *infinite*, by showing that it contains a sequence $s_1, \ldots, s_n, \ldots$ of distinct points. To begin with, there is, by hypothesis, some formula A and some point s_1 such that $\mathcal{M} \models_{s_1} \Box A \wedge \neg\Box^2 A$.

Now make the inductive assumption that s_n has been defined and has

$$\mathcal{M} \models_{s_n} \Box^n A \wedge \neg\Box^{n+1} A. \qquad (\P)$$

But the formula

$$\Box^n A \wedge \neg\Box^{n+1} A \rightarrow \Diamond(\Box^{n+1} A \wedge \neg\Box^{n+2} A)$$

is (equivalent to) an instance of the schema Mk, so as $\mathcal{M} \models \Lambda_*$, from (¶) it follows that there is some point s_{n+1} with $s_n R s_{n+1}$ and

$$\mathcal{M} \models_{s_{n+1}} \Box^{n+1} A \wedge \neg\Box^{n+2} A.$$

Hence, by induction, s_n is defined to satisfy (¶) for all positive integers n. But then to see that the s_n's are distinct, observe that if $n < m$, then

$\Box^{n+1}A$ is false at s_n by (¶), but true at s_m since $\Box^m A$ is true at s_m, and $\mathcal{M} \models \Box^m A \rightarrow \Box^{n+1}A$ from the schema T.

Proof of (2). This requires the construction of a Λ_*-model that rejects schema 4. In view of (1), this model will have to be infinite. Let

$$\mathcal{F}_r = (\omega, R),$$

where ω is the set $\{0, 1, 2, \ldots\}$ of natural numbers, and mRn iff $m \leq n+1$, so that each number is R-related to all numbers big than or equal to its predecessor. $\mathcal{F}_r$, which first appeared in Makinson [1969], has become known as *the recession frame.*

Exercises 7.8

(1) Show that $\mathcal{F}_r$ validates the logic Λ_*.

(2) Show that $\mathcal{F}_r$ is not transitive, and so carries a model in which an instance of schema 4 is false. Hence complete the argument showing that Λ_* lacks the finite model property.

Incompleteness

The canonical model construction shows that any consistent normal logic is determined by some *model.* On the other hand, there are consistent logics that are not determined by any class of *frames.* The first example of such an *incomplete* logic was a temporal one, discovered by Thomason [1972]. It can be defined as the smallest temporal logic Λ_T containing the schemata

$$\begin{array}{ll} L_{\mathrm{F}}: & [\mathrm{F}](A \wedge [\mathrm{F}]A \rightarrow B) \vee [\mathrm{F}](B \wedge [\mathrm{F}]B \rightarrow A) \\ D_{\mathrm{F}}: & <\mathrm{F}> \top \\ W_{\mathrm{P}}: & <\mathrm{P}> A \rightarrow <\mathrm{P}> (A \wedge \neg <\mathrm{P}> A) \\ M_{\mathrm{F}}: & [\mathrm{F}] <\mathrm{F}> A \rightarrow <\mathrm{F}> [\mathrm{F}]A. \end{array}$$

The import of M_{F} is that it requires the truth-value of A to eventually "settle down" to a fixed value. For, if the antecedent is true, then at any future time there will be a time after that at which A is true. But if the consequent is false, then at any future time there will be a time after that at which A is false. Thus if M_{F} is to be true, then A must eventually become either true forever or false forever.

It turns out that Λ_T has no frames at all! To see this, observe first that any time-frame validating W_{P} is *irreflexive*, for if sRs, then putting $V(A) = \{s\}$ for an atomic A would falsify W_{P} at s. Thus if $\mathcal{F} \models \Lambda_T$, then R is weakly future-connected, by L_{F}, so for any point s the set $X_s = \{t : sRt\}$ is a strict total ordering (connected, irreflexive) which, by D_{F}, has no last element. But then we can choose a subset Y of X_s such that neither Y nor $X_s - Y$ has a last element. Putting $V(A) = Y$ then gives a model in which M_F is false at s. However this contradicts the hypothesis that $\mathcal{F} \models \Lambda_T$.

To see that Λ_T is nonetheless consistent, it suffices to construct a *model* for it (by the argument just given, the frame of this model must carry other models that falsify Λ_T). Let $\mathcal{M} = (\omega, <, V)$, where $\omega = \{0, 1, 2, \ldots\}$, and $V(p) = \emptyset$ for all $p \in \Phi$. The frame $(\omega, <)$ validates all axioms of Λ_T except M_F. An inductive argument shows that for any formula A, the set $\{n \in \omega : \mathcal{M} \models_n A\}$ is either finite, or *cofinite* (i.e. has a finite complement). Thus "as time passes", A eventually becomes either false forever (finite case), or true forever (cofinite case). In the first case $[\mathrm{F}] \langle \mathrm{F} \rangle A$ is false everywhere, and in the second case $\langle \mathrm{F} \rangle [\mathrm{F}] A$ is true everywhere. Hence $\mathcal{M} \models M_F$.

Exercise 7.9

Fill in all the details of the above argument.

Incomplete $\Box$-Logics

After the discovery of Λ_T, a number of incomplete logics in the language of a single modal connective $\Box$ were produced (Thomason [1974], Fine [1974], van Benthem [1978]). The latest, and seemingly simplest, example appears in a paper by Boolos and Sambin [1985], where its discovery is attributed to R. Magari. The logic is KH, where H is the schema

$$\Box(\Box A \leftrightarrow A) \rightarrow \Box A.$$

Notice that $KH \subseteq KW$, where W, as above, is

$$\Box(\Box A \rightarrow A) \rightarrow \Box A.$$

We noted in §1 that any frame for W is transitive (Boolos [1979], p.82), and hence validates

$$4: \quad \Box A \rightarrow \Box\Box A.$$

Boolos and Sambin show that H and W are valid on exactly the same frames, implying that any KH-frame must validate 4. They then give a model for H in which 4 is false, showing that 4 is not KH-deducible.

To spell out some details, suppose $\mathcal{F} \models H$. To prove $\mathcal{F}$ is transitive, take a point s in order to show that

$$sRt \;\&\; tRu \quad \text{implies} \quad sRu.$$

Let $\mathcal{M}$ be any model on $\mathcal{F}$ in which

$$V(p) = \{t : u \in S^t \text{ implies } sRu\}$$

(recall that $S^t = \{v : tR^*v\}$ is the subframe generated by t).

Exercise 7.10

Show that $\mathcal{M} \models_s \Box(\Box p \leftrightarrow p)$.

Since $\mathcal{F} \models H$, it follows from this exercise that $\mathcal{M} \models_s \Box p$. Hence if sRt and tRu, we have $t \in V(p)$ and $u \in S^t$, so sRu as desired.

The following intransitive H-model is due to M. J. Cresswell. It is an extension of the construction of Makinson's recession frame $\mathcal{F}_r$, and is just one of a number of uses to which $\mathcal{F}_r$ has been put in studies of the "pathology" of modal logics (cf. Bull and Segerberg [1984], §19; Hughes and Cresswell [1984]). Most spectacularly, $\mathcal{F}_r$ was used by Blok [1980] to prove that if Λ is any normal extension of KT, there are *uncountably* many other logics having exactly the same frames as Λ! All but one of these uncountably many logics must be incomplete.

Let $\mathcal{M} = (\mathbb{Z}, R, V)$, where $V(p) = \mathbb{Z} - \{0\}$ for all $p \in \Phi$, and R is a nonstandard ordering of the points of $\mathbb{Z}$ got by shifting all the negative integers "to the right",

$$0, 1, 2, \ldots, n, \ldots\ldots \quad \ldots\ldots, -n, \ldots, -2, -1$$

and then allowing each non-negative integer to also have itself and its predecessor as R-alternatives (hence destroying transitivity). Formally, if $m, n \in \mathbb{Z}$, then mRn iff one of the following hold.

$$n < 0 \leq m$$
$$0 \leq m \leq n + 1$$
$$m < n < 0.$$

Exercises 7.11

(1) Show that $\Box p \to \Box\Box p$ is false at the point 2 in $\mathcal{M}$.

(2) Show that for all formulae A, the set $\{m : \mathcal{M} \models_m A\}$ is either finite or cofinite. Use this to prove $\mathcal{M} \models H$.

In conclusion, note that the axiom W_{P} in Thomason's logic Λ_T is a variant of the schema

$$[\mathrm{P}]([\mathrm{P}]A \to A) \to [\mathrm{P}]A,$$

which is the past-tense version of the schema W. The latter has manifestly played an important role in technical studies of the metatheory of modal logics. There is another context in which it is also important: KW is the logic that results when $\Box$ is interpreted as meaning "it is provable in Peano Arithmetic that". This is explained in the book by Boolos [1979].

Undecidability

A logic with the finite frame property is decidable, *provided that it is finitely axiomatisable.* This last qualification is essential: there exist logics with the finite frame property that are undecidable. In fact, Urquhart [1981] showed that for any subset X of ω there exists a logic Λ_X with the finite frame property, such that Λ_X has the same "degree of unsolvability" as X. We now discuss this result, using the following definitions.

- A point s in a frame (S, R) is *dead* if there is no $t \in S$ with sRt.
- A point is *live* if it is not dead.
- For $n \in \omega$, $\mathcal{F}_n$ is the frame depicted as

$$-1 \leftarrow 0 \rightarrow 1 \rightarrow \cdots \rightarrow n,$$

  i.e. $\mathcal{F}_n = (\{-1, 0, 1, \ldots, n\}, R)$, with

$$R = \{\langle 0, -1\rangle\} \cup \{\langle m, m+1\rangle : 0 \leq m < n\}.$$

- $\mathcal{G}_n$ is the frame depicted

$$0 \rightarrow 1 \rightarrow \cdots \rightarrow n,$$

  which results by removing the point -1 from $\mathcal{F}_n$.
- $\widehat{\mathbf{0}}$ is the formula $\Diamond\Box\bot \wedge \Diamond^2\top$.
- A_n is the formula $\widehat{\mathbf{0}} \rightarrow \Box^{n+2}\Diamond\top$.

Note that if $n \geq 2$, then 0 is distinguished in $\mathcal{F}_n$ as the only point that is R-related both to a dead point and to a live one. This accounts for the superscript "$n+2$" in the definition of A_n, and the emphasis on the frames $\mathcal{F}_{n+2}$ in what follows.

Exercises 7.12

(1) In any model,
$\mathcal{M} \models_s \Box\bot$ iff s is dead;
$\mathcal{M} \models_s \Diamond\top$ iff s is live.

(2) For any $n \in \omega$, $\mathcal{F}_{n+2} \models_s \widehat{\mathbf{0}}$ iff $s = 0$. (Note: since $\widehat{\mathbf{0}}$ contains no atomic formulae, its truth at any point in $\mathcal{F}_{n+2}$ is independent of any particular model on that frame.)

(3) $\mathcal{F}_{n+2} \not\models_0 A_n$

(4) $\mathcal{F}_{n+2} \models A_j$ if $j \neq n$.

(5) $\mathcal{G}_n \models \neg\widehat{\mathbf{0}}$, and hence $\mathcal{G}_n \models A_j$ for all j.

Now let X be an arbitrary set of natural numbers. Put

$$\mathcal{C}_X = \{\mathcal{G}_n : n \in \omega\} \cup \{\mathcal{F}_{n+2} : n \notin X\},$$

and let $\Lambda_X = \{B : \mathcal{C}_X \models B\}$ be the logic determined by $\mathcal{C}_X$. Since all members of $\mathcal{C}_X$ are finite, it is immediate that Λ_X has the finite frame property.

Lemma 7.13. *For any $j \in \omega$,*

$$\vdash_{\Lambda_X} A_j \quad \text{iff} \quad j \in X.$$

Proof. Suppose $j \in X$. Then if $n \notin X$, $j \neq n$, so $\mathcal{F}_{n+2} \models A_j$ by 7.12(4). Together with 7.12(5), this shows that $\mathcal{C}_X \models A_j$, as desired.

On the other hand, if $j \notin X$, then $\mathcal{F}_{j+2} \in \mathcal{C}_X$ and so by 7.12(3), $\mathcal{C}_X \not\models A_j$.

Corollary 7.14. *If X is undecidable, then so is Λ_X.*

Proof. Since formula A_j is explicitly defined in terms of j, the Lemma shows that a decision procedure for theoremhood in Λ_X would yield a decision procedure for membership of X.

Axioms for Λ_X

We now develop an axiomatisation for Λ_X, and strengthen the analysis to prove Urquhart's result that there is an undecidable Λ_X that has a decidable set of axioms. We need the following schemata

Pfun:	$\Box(\Diamond A \to \Box A)$
De:	$\Diamond(\Box\bot \wedge A) \to \Box(\Box\bot \to A)$
Li:	$\Diamond(\Diamond\top \wedge A) \to \Box(\Diamond\top \to A)$

Exercises 7.15

Let Λ be any normal logic containing *Pfun*, *De*, and *Li*. Work in the canonical model $\mathcal{M}^\Lambda$ for Λ.

(1) Use *Pfun* to show that if $sR^\Lambda t$, then t itself is R^Λ-related to at most one point in S^Λ.

(2) Use *De* to show that each $s \in S^\Lambda$ is R^Λ-related to at most one dead point.

(3) Use *Li* to show that each $s \in S^\Lambda$ is R^Λ-related to at most one live point.

Theorem 7.16. *If $\omega - X$ is infinite, then Λ_X is the smallest normal logic that contains $Pfun$, De, Li, and $\{A_j : j \in X\}$.*

Proof. Let Λ be the smallest normal logic that contains $Pfun$, De, Li, and $\{A_j : j \in X\}$. Now $Pfun$, De and Li are all valid in all frames of the type $\mathcal{G}_n$ and $\mathcal{F}_{n+2}$, hence in $\mathcal{C}_X$, so it follows from Lemma 7.13 that $\Lambda \subseteq \Lambda_X$.

To prove the converse, suppose $\nvdash_\Lambda A$, with the objective of showing $\nvdash_{\Lambda_X} A$, i.e. $\mathcal{C}_X \not\models A$. Let s_0 be a point in $\mathcal{M}^\Lambda$ at which A is false, and let $\mathcal{M} = (S, R, V)$ be the submodel of $\mathcal{M}^\Lambda$ generated by s_0, whence $\mathcal{M} \not\models_{s_0} A$.

Since each point is either live or dead, (2) and (3) of 7.15 imply that s_0 is R-related to at most *two* points.

Case 1: s_0 has at most one R-related point. If there is one, say s_1, then by 7.15(1), this will in turn be R-related to at most one point, and so on. Hence $\mathcal{M}$ takes the form of an "R-sequence"

$$s_0 R s_1 R \cdots R s_n \cdots\cdots \qquad (\%)$$

Now if some s_n is dead, then the sequence stops at s_n, and the frame of $\mathcal{M}$ is isomorphic to $\mathcal{G}_n$. Since $\mathcal{G}_n \in \mathcal{C}_X$, this yields the desired result $\nvdash_{\Lambda_X} A$.

If however all terms of the sequence are live, there are two further possibilities. First, it may be that all terms are distinct. In that case we can truncate the sequence at some point that bounds the "degree of nesting" of $\Box$ in A, and get a finite falsifying model for A. To make this idea more precise, the *modal degree* of an arbitrary formula B is introduced as the number $deg(B)$ defined inductively by

$$deg(p) = deg(\bot) = 0,$$
$$deg(B_1 \to B_2) = \text{maximum of } deg(B_1) \text{ and } deg(B_2),$$
$$deg(\Box B) = deg(B) + 1.$$

Then to determine the truth-value of A at s_0 we need not "look beyond" s_n in $\mathcal{M}$ provided that $n > deg(A)$, as the reader may verify. Taking such an n, and restricting to $s_0 R s_1 R \cdots R s_n$ again gives a falsifying model for A on a frame isomorphic to $\mathcal{G}_n$.

Finally, if the terms are not all distinct, then there must be some s_n such that $s_n R s_k$ for some (unique) $k < n$. Then we pick the least such n and "unravel the loop": the sequence up to s_n is taken, and extended by appending a new copy $s'_k \cdots s'_n$ of the segment from s_k to s_n, making truth-values at s'_{k+i} agree with those at s_{k+i}. Further new copies of this segment are appended until a sequence is built that is longer than $deg(A)$. Again this produces a falsifying model for A on a frame isomorphic to some $\mathcal{G}_m$.

Case 2: s_0 is R-related to two points. Then one, say s_{-1}, is dead, and the other, s_1, is live, so that there is some s_2 with $s_1 R s_2$. Moreover, s_2 in turn can be R-related to at most one point, and so on, so that $\mathcal{M}$ consists of s_{-1} and an R-sequence of the form (%) above.

This R-sequence is then analysed as in Case 1. If some s_{n+2} is dead, then the frame of $\mathcal{M}$ is isomorphic to $\mathcal{F}_{n+2}$, unless we have $s_{n+2} = s_{-1}$. But in the latter case there is an evident p-morphism from $\mathcal{F}_{n+2}$ onto the frame of $\mathcal{M}$ in which m maps to s_m for $-1 \leq m < n+2$, and $n+2$ maps to the dead point s_{-1}. In either event we are led to a falsifying model for A on $\mathcal{F}_{n+2}$. But since $\Diamond\top \notin s_{n+2}$ and $\widehat{\mathbf{0}} \in s_0$, it follows that $A_n \notin s_0$, hence $\not\vdash_\Lambda A_n$, and so $n \notin X$. This implies that $\mathcal{F}_{n+2} \in \mathcal{C}_X$, giving $\mathcal{C}_X \not\models A$ as desired.

If no s_{n+2} is dead, then we adapt the sequence (%) by truncation and/or extension as in Case 1, to again obtain a falsifying model for A on a frame isomorphic to some $\mathcal{F}_{n+2}$. Here, finally, we invoke the assumption that $\omega - X$ is infinite, for then we can choose to take n large enough that $n \notin X$ as well as $n+2 > deg(A)$. Again this gives $\mathcal{F}_{n+2} \in \mathcal{C}_X$.

This completes the proof of Theorem 7.16.

Corollary 7.17. *There exists an undecidable logic that has a decidable set of axioms.*

Proof. Let X be an effectively enumerable but undecidable set of natural numbers. Then Λ_X is undecidable, and $\omega - X$ is infinite, so the axioms for Λ_X are as described in Theorem 7.16. Since X is effectively enumerable, so too are these axioms. But by a well known metalogical result due to Craig, a logic with an effectively enumerable set of axioms has a decidable such set. Indeed, if $\Delta = \{A_0, A_1, \dots, A_n, \dots\}$ is an effectively enumerable set of formulae, then the smallest logic containing Δ is also the smallest logic containing the decidable set

$$\{A_0, A_0 \wedge A_1, \dots, A_0 \wedge \cdots \wedge A_n, \dots\dots\}.$$

To conclude the discussion of the logics Λ_X, we observe that if $\omega - X$ is not infinite, a simpler axiomatisation is obtained: only finitely many axioms are needed.

Theorem 7.18. *If $\omega - X = \{n_1, \dots, n_k\}$, then Λ_X is the smallest normal logic that contains $Pfun$, De, Li, and the formula*

$$A_X: \quad \widehat{\mathbf{0}} \to \Diamond^{n_1+2}\Box\bot \vee \cdots \vee \Diamond^{n_k+2}\Box\bot.$$

Proof. In this case, Λ_X is the logic determined by

$$\mathcal{C}_X = \{\mathcal{G}_n : n \in \omega\} \cup \{\mathcal{F}_{n_1+2}, \dots, \mathcal{F}_{n_k+2}\}.$$

Following the proof of Theorem 7.16, when it comes to Case 2 the formulae $\widehat{\mathbf{0}}$ and A_X are both in s_0, and hence so is $\Diamond^{n_i+2}\Box\bot$ for some $1 \leq i \leq k$. But this is only possible if s_{n_i+2} is dead, leading to a falsifying model for A on $\mathcal{F}_{n_i+2}$.

Part Two

Some Temporal and Computational Logics

8 | Logics with Linear Frames

Part Two applies the techniques developed in the previous sections to some standard temporal logics, and to some modal logics that have been employed in the theory of computation. One of these, examined in §9, involves the use of temporal logic to express properties of linear state sequences generated by concurrent programs. To study this logic, it is helpful to first isolate its "[F]-fragment", and axiomatise the logic determined by the frame $(\omega, \leq)$ in the language of a single modal connective. This will be done in the present section, in the context of a general study of discrete, dense, and continuous time. §10 introduces the multi-modal language of *dynamic* logic, in which the modal connectives are indexed by the commands of a programming language.

Discrete Future Time

In the $\Box$-language of §1, let Ω be the logic $K4DLZ$, with axioms

$$\begin{array}{ll} 4: & \Box A \to \Box\Box A \\ D: & \Box A \to \Diamond A \\ L: & \Box(A \wedge \Box A \to B) \vee \Box(B \wedge \Box B \to A) \\ Z: & \Box(\Box A \to A) \to (\Diamond\Box A \to \Box A). \end{array}$$

Our first goal in this section is to prove that

$$\vdash_\Omega A \quad \text{iff} \quad (\omega, <) \models A.$$

Each of the axioms of Ω embodies a feature of the frame $(\omega, <)$. 4 corresponds to transitivity, D to seriality ("endless time"), and L to connectedness. Z embodies an aspect of the *discreteness* of $(\omega, <)$, namely that between any two points there are only finitely many other points. What this has to do with Z may be learned from

Exercises 8.1

(1) Show that $(\omega, <) \models Z$.

(2) *Soundness* of Ω : Prove that

$$\vdash_\Omega A \quad \text{implies} \quad (\omega, <) \models A.$$

(3) Let $\mathcal{F} = (\omega \cup \{\infty\}, R)$, with mRn iff $m < n \in \omega$ or $n = \infty$. Show that $\mathcal{F} \not\models Z$.

Clusters

In order to carry through a completeness theorem for Ω, we need a fine analysis of the structure of any *transitive* frame $\mathcal{F} = (S, R)$. In such a frame, define an equivalence relation $\approx$ on S by

$$s \approx t \quad \text{iff} \quad s = t \text{ or } (sRt \text{ and } tRs).$$

The equivalence class

$$C_s = \{t : s \approx t\}$$

is called the *R-cluster* of s.

Ordering Clusters

Putting

$$C_s \leq C_t \quad \text{iff} \quad sRt$$

gives a well-defined relation between clusters that is transitive and anti-symmetric. Hence putting

$$\begin{aligned} C_s < C_t \quad &\text{iff} \quad C_s \leq C_t \ \& \ C_s \neq C_t \\ &\text{iff} \quad sRt \text{ and not } tRs, \end{aligned}$$

defines $<$ to be a strict ordering, i.e. transitive and irreflexive, therefore asymmetric.

We tend to use evocative terminology in reference to clusters, so that if $C < C'$ we might say that cluster C' is *later than*, or *comes after*, cluster C, while C *precedes*, or is *earlier than*, or *comes before*, C', and so on.

Exercises 8.2

(1) Verify that $\approx$ is an equivalence relation.

(2) Verify that $\leq$ and $<$ are well-defined relations, and have the asserted properties.

(3) In a transitive model, if $C_s = C_t$, show that for any formula B,

$$\mathcal{M} \models_s \Box B \quad \text{iff} \quad \mathcal{M} \models_t \Box B.$$

Types of Cluster

If a cluster C contains more than one point, then the relation R is reflexive in C. For if $C = C_s = C_t$, with $s \neq t$, then since $s \approx t$, we have sRt and tRs, and so sRs by transitivity. Hence in particular, $C \leq C$.

A cluster C is *degenerate* if $C \not\leq C$. In view of the foregoing, this must mean that C consists of a single irreflexive point: $C = \{s\}$, with not sRs. Notice that all clusters in a strict total ordering, like $(\omega, <)$, are degenerate.

We distinguish two types of *non-degenerate* (i.e. $C \leq C$) cluster. A *simple* cluster consists of a single reflexive point: $C = \{s\}$ with sRs. All clusters in the frame $(\omega, \leq)$ are simple. A *proper* cluster is a (non-degenerate) cluster with at least two points. Observe that the relation R is *universal* on any non-degenerate cluster.

If R is connected, i.e.

$$\forall s \forall t(sRt \vee s = t \vee tRs),$$

then $<$ is a strict total ordering of clusters, and so $\mathcal{F}$ takes the form of a sequence of clusters, as illustrated in the following diagram, where the bullets $\bullet$ depict degenerate clusters, and the circles are non-degenerate ones.

$$\cdots \to \bigcirc \to \bullet \to \bigcirc \to \bullet \to \bullet \to \bigcirc \to \cdots$$

If S is finite, then this sequence will have a *first* and a *last* cluster.

Balloons

A *balloon* is a finite transitive and connected frame whose last cluster is non-degenerate, while all other clusters are degenerate:

$$\bullet \to \bullet \to \cdots\cdots \to \bullet \to \bigcirc$$

(there need not actually be any non-last clusters, so a finite universal frame, comprising a single non-degenerate cluster, is counted as a balloon).

Exercises 8.3

(1) If $\mathcal{F}$ is a balloon, show that $\mathcal{F} \models \Omega$.

(2) If $\mathcal{F}$ is a transitive frame that has a *non-degenerate* cluster C that is not last (i.e. $C < C'$ for some cluster C'), show that $\mathcal{F} \not\models Z$.

Theorem 8.4. *If $\mathcal{F}$ is a balloon, then $\mathcal{F}$ is a p-morphic image of $(\omega, <)$.*

Proof. Let $S = \{s_0, \ldots, s_{n-1}, t_0, \ldots, t_{m-1}\}$, where $\{s_0\}, \ldots, \{s_{n-1}\}$ are the degenerate clusters in $<$-order (if there are any), and $\{t_0, \ldots, t_{m-1}\}$ is the non-degenerate last cluster. Define $f : \omega \to S$ by

$$\begin{aligned} f(i) &= s_i & (0 \leq i < n) \\ f(n + q \cdot m + j) &= t_j & (0 \leq j < m,\ q \in \omega). \end{aligned}$$

As a sequence, f looks like

$$s_0, \ldots, s_{n-1}, t_0, \ldots, t_{m-1}, t_0, \ldots, t_{m-1}, t_0, \ldots, t_{m-1}, \ldots\ldots$$

with the last cluster repeated forever. Since R is universal on this last cluster, the properties of a p-morphism are satisfied, as the reader should verify.

Completeness of $K4DLZ$

lt follows from Theorem 8.4 that if $(\omega, <) \models A$, then A is valid in all balloons. Thus to prove that the logic Ω is complete with respect to $(\omega, <)$, it suffices to show that it is complete with respect to the class of balloons, i.e. that any non-theorem of Ω is falsified by a model on some balloon. This will also show that Ω has the finite frame property, and so is decidable (cf. Theorem 4.7 and Exercise 4.8(2)).

Suppose then that $\nvdash_\Omega A$. As just explained, we want to find a balloon in which A is not valid.

First Model. Since $\nvdash_\Omega A$, A is false at some point s_A in the canonical model $\mathcal{M}^\Omega$. In virtue of the schemata 4, D, and L, $\mathcal{M}^\Omega$ is transitive, serial, and weakly connected.

Second Model. Let $\mathcal{M} = (S, R, V)$ be the submodel of $\mathcal{M}^\Omega$ generated by s_A. Then by the Submodel Lemma 1.7, A is false at s_A in $\mathcal{M}$, and $\mathcal{M} \models \Omega$ because $\mathcal{M}^\Omega \models \Omega$.

Also, R is transitive, serial, and *connected* (Exercise 3.11(1)).

Third Model. Let $\Gamma = Sf(A)$, and let

$$\mathcal{M}^\tau = (S_\Gamma, R^\tau, V_\Gamma)$$

be the *transitive* Γ-filtration of $\mathcal{M}$ (Exercise 4.5(3)). By the Filtration Lemma 4.3, A is false at $|s_A|$ in $\mathcal{M}^\tau$. Also S_Γ is finite (4.1), while R^τ is transitive, serial, and connected (4.5(5)).

Thus the frame of $\mathcal{M}^\tau$ consists of a finite sequence of clusters. Moreover, the last cluster, C_x say, is non-degenerate. For, by seriality, there is some y with $xR^\tau y$, and so $C_x \leq C_y$. But then $C_x = C_y$, as C_x is last, making $C_x \leq C_x$.

However, at this point we cannot rule out the possibility that $\mathcal{M}^\tau$ has some non-degenerate cluster that is not last, so that the frame of $\mathcal{M}^\tau$ is not a balloon. Hence $\mathcal{M}^\tau$ may not be the model we are seeking.

Fourth Model. (Balloon Surgery)

If $\mathcal{M}^\tau$ does have a non-last cluster C that is non-degenerate, then we could try to remove it by weakening the relation R^τ in C to some strict total ordering, so that C is replaced by a sequence of degenerate clusters. We would want to do this in such a way that the truth-values of members of Γ were left unchanged, so that our non-Ω-theorem A is still false at $|s_A|$ in the new model.

The potential problem with this idea is that formulae of the form $\Box B$ that are false at certain points in C in $\mathcal{M}^\tau$ may cease to be false, because the R^τ-alternative at which B is false may no longer be an alternative in the new model. However this turns out not to be a problem in the presence of the axiom schema Z, which is true in the second model $\mathcal{M}$.

Z-Lemma 8.5. *Let $C_{|s|}$ be a non-last R^τ-cluster. Then if $\Box B \in \Gamma$ and $\mathcal{M} \not\models_s \Box B$, there exists a point $t \in S$ with $\mathcal{M} \not\models_t B$ and $C_{|s|} < C_{|t|}$.*

Proof. Let $\Box B$ be in Γ and false at s in $\mathcal{M}$.

Case 1. Suppose $\mathcal{M} \models_s \Diamond\Box B$. Then since $\mathcal{M} \models Z$,

$$\mathcal{M} \not\models_s \Box(\Box B \to B),$$

so there exists $t \in S$ with sRt, $\mathcal{M} \models_t \Box B$, and $\mathcal{M} \not\models_t B$. Then $|s|R^\tau|t|$, since sRt and R^τ is a Γ-filtration of R. But since $\Box B$ is true at t and false at s in $\mathcal{M}$, the definition of R^τ implies that we do not have $|t|R^\tau|s|$. Hence the cluster of $|t|$ comes strictly *after* that of $|s|$.

Case 2. Suppose instead that $\mathcal{M} \not\models_s \Diamond\Box B$. Now since $C_{|s|}$ is not last, there exists $u \in S$ with $C_{|s|} < C_{|u|}$. Then we cannot have uRs or $u = s$, or else $|u|R^\tau|s|$ or $|u| = |s|$, making $C_{|u|} \leq C_{|s|}$. Hence, *as R is connected*, sRu. Then in $\mathcal{M}$, since $\Diamond\Box B$ is false at s, $\Box B$ is false at u, so B is false at some t with uRt. We have $C_{|s|} < C_{|u|} \leq C_{|t|}$, and so the Z-Lemma is proved.

Final Model

For each non-degenerate non-last cluster C of $\mathcal{M}^\tau$, let $<_C$ be a strict total ordering of the points of C. Define

$$\mathcal{M}' = (S_\Gamma, R', V_\Gamma),$$

where $xR'y$ holds if and only if $xR^\tau y$ and either x and y do not belong to the same non-degenerate non-last cluster, or else $x <_C y$ for some such cluster C. Then the frame of $\mathcal{M}'$ is a balloon. For each $B \in \Gamma$ and $s \in S$ we have

$$\mathcal{M} \models_s B \quad \text{iff} \quad \mathcal{M}' \models_{|s|} B. \tag{†}$$

This is proven by induction on the formation of B, using the fact that R' is contained in R^τ and so satisfies the second filtration condition. The only problematic case in the proof is taken care of by the Z-Lemma.

It follows in particular that $\mathcal{M}' \not\models_{|s_A|} A$, so we have found our falsifying model on a balloon for the non-Ω-theorem A. Since S_Γ has at most 2^n elements, where n is the number of subformulae of A, we also get the strong finite frame property for the logic Ω.

Exercise 8.6

Work through the proof of (†).

Completeness for KW

The schema Z is weaker than the ubiquitous

$$W: \qquad \Box(\Box A \to A) \to \Box A,$$

and if we had $\mathcal{M} \models W$ in the proof of the Z-Lemma, Case 2 would become redundant, therefore so too would the hypotheses that R is connected and $C_{|s|}$ is not last. From this observation, a completeness proof for KW readily emerges:

Exercises 8.7

(1) Prove that KW is determined by the class of finite strict orderings, and is decidable (remember that $KW = K4W$).

(2) (*Alternative completeness proof.*) If Γ is a finite set of formulae closed under subformulae, and $\mathcal{M}$ is the canonical model of a normal logic containing KW, define

$$\mathcal{M}' = (S_\Gamma, R', V_\Gamma),$$

where

$$xR'y \quad \text{iff} \quad xR^\tau y \text{ and not } yR^\tau y.$$

Prove that R' is a strict ordering, and that

$$\mathcal{M} \models_s B \quad \text{iff} \quad \mathcal{M}' \models_{|s|} B$$

for all $B \in \Gamma$. Use this to obtain the results of Exercise (1).

Dense and Continuous Time

It was claimed in §6 that the real-number and rational-number frames $(\mathbb{R}, <)$ and $(\mathbb{Q}, <)$ determine the same logic. This logic is $K4DLX$, where X is the schema

$$\Box\Box A \to \Box A,$$

corresponding to the condition of weak density on R (Theorems 1.12, 1.13, 3.6). The following exercises show how to prove these determination results, and also the corresponding results for reflexive time. The latter involve the logic $S4.3$.

Exercises 8.8

Let P be either $\mathbb{R}$ or $\mathbb{Q}$. A *right-open* interval in P is a subset of P having one of the forms

$$[r, q) = \{p \in P : r \leq p < q\}, \qquad (r, q) = \{p \in P : r < p < q\},$$

for some r and q. In each case, q is the *right end-point* of the interval. We allow $q = \infty$ here, with, as usual,

$$[r, \infty) = \{p \in P : r \leq p\} \quad \text{and} \quad (r, \infty) = \{p \in P : r < p\}.$$

Observe that by the density of $<$ in P, any right-open interval can be decomposed as the disjoint union of n right-open intervals, for any positive integer n.

Next, let $\mathcal{M}$ be a generated submodel of the canonical model for $K4DLX$, and $\mathcal{M}^\tau$ the transitive Γ-filtration of $\mathcal{M}$ for a suitable finite Γ, as in the completeness proof for $K4DLZ$.

(1) Show that for any right-open interval I in P, the frame $(I, <)$ can be mapped p-morphically onto any *non-degenerate* cluster in $\mathcal{M}^\tau$ (hint: choose in the interval a strictly increasing sequence that converges to the right end-point, and adapt the construction of Theorem 8.4).

(2) Show that $\mathcal{M}^\tau$ does not contain any *adjacent* degenerate clusters, i.e. any degenerate cluster is immediately followed by a non-degenerate one (hint: this uses connectedness as well as weak density of R).

(3) Let I be a right-open interval in $(P, <)$ of the form $[r, q)$. Apply the previous two exercises to show that there is a p-morphism from $(I, <)$ onto the frame of $\mathcal{M}^\tau$, by mapping appropriate subintervals onto non-degenerate clusters, and the right end-points of intervals onto any degenerate clusters that may be present. Deduce that

$$(I, <) \models A \quad \text{iff} \quad \vdash_{K4DLX} A.$$

(4) Use the Submodel Lemma 1.7 to show that the determination result of Exercise 3 can be extended to hold for $I = (r, q)$ for any r, including $r = \infty$, and in particular for $I = \mathbb{R}$ and $I = \mathbb{Q}$.

(5) Adapt the above constructions to show that if I is a right-open interval in P, or any of the sets listed in the previous exercise, then

$$(I, \leq) \models A \quad \text{iff} \quad \vdash_{S4.3} A.$$

(6) Having worked through the foregoing, it should be becoming clear just what properties of a total ordering suffice for it to determine one of the logics $K4DLX$ and $S4.3$. Write down a minimal list of properties that suffice in each case.

The Discrete Diodorean Case

The logic determined by the reflexive frame $(\omega, \leq)$ is $S4.3Dum$, that is to say $KT4LDum$, where the schema Dum (named for Michael Dummett), is

$$\Box(\Box(A \to \Box A) \to A) \to (\Diamond\Box A \to A).$$

Exercise 8.9

Show that *Dum* is valid in $(\omega, \leq)$.

Cluster Analysis for Extensions of $S4.3$

To prove the claimed completeness result for $S4.3Dum$ we follow the general strategy of the completeness theorem for Ω, but find that we have to make a more refined analysis of the structure of clusters, using a certain connected subrelation of R^{τ}. This analysis will also be used in the discussion to follow of Bull's Theorem about extensions of S4.3, and in our study of temporal program logics in §9.

So, let $\mathcal{M} = (S, R, V)$ be a generated submodel of the canonical model of a normal logic that contains $S4.3$. Then R is reflexive, transitive, and connected. Let Γ be a finite subformula-closed set, as usual, and define a relation R^c on S_Γ by

$$xR^c y \quad \text{iff} \quad \forall s \in x \, \exists t \in y \, (sRt).$$

Lemma 8.10.
(1) *R^c is reflexive, transitive, and connected.*
(2) *R^c is contained in any Γ-filtration of R.*

Proof.

(1). Reflexivity and transitivity of R^c follow readily from the fact that R has these properties. For connectivity, suppose that it is not the case that $xR^c y$. Then there exists $s \in x$ such that sRt fails for all $t \in y$. But then tRs holds for all $t \in y$, since R is connected. This shows that $yR^c x$.

(2). Suppose $xR^c y$. Take any $s \in x$. Then sRt for some $t \in y$. But if R' is any Γ-filtration of R, sRt implies $|s|R'|t|$, i.e. $xR'y$, as desired.

Clusters Within Clusters

The first part of Lemma 8.10 implies that the frame (S_Γ, R^c) consists of a finite sequence of R^c-clusters (recall that for reflexive relations, there are no degenerate clusters). From the second part we see, in particular, that $R^c \subseteq R^{\tau}$, where R^{τ} is the transitive Γ-filtration of R, and so the R^c-cluster of any point is entirely contained within the R^{τ}-cluster of that point. Thus each R^{τ}-cluster itself decomposes into a sequence of R^c-clusters. Thus if C is a given R^{τ}-cluster, there is within C an R^c-cluster C^H that comes after all the other R^c-clusters that are included in C. C^H will be called the *head* of C. The situation is depicted in the following diagram, in which the rectangles represent R^{τ}-clusters, the circles are R^c-clusters, and a circle with a dot in the centre is the head of its R^{τ}-cluster.

$$\cdots \to \boxed{\bigcirc \to \cdots \to \odot} \to \boxed{\bigcirc \to \cdots \to \odot} \to \cdots$$

R^c-Lemma 8.11. *If $\Box B \in \Gamma$ and $\mathcal{M} \not\models_s \Box B$, then there exists t with $\mathcal{M} \not\models_t B$ and $|s|R^c|t|$. Moreover, if $|s|$ is not in the head of its R^τ-cluster, then not $|t|R^c|s|$, i.e. $|t|$ is in a later R^c-cluster than $|s|$.*

Proof. Let C be the R^τ-cluster of $|s|$, and let z be a member of the head C^H of C. Then $|s|R^c z$.

Next, let X be the union of all the R^c-clusters that precede the R^c-cluster C^H, i.e.

$$X = \{x \in S_\Gamma : xR^c z \text{ and not } zR^c x\}.$$

Then if $X = \{x_0, \dots, x_m\}$, for each $j \leq m$ we have not $zR^c x_j$, and so there exists $s_j \in z$ such that not $s_j Rt$ for all $t \in x_j$.

Now as R is connected, the s_j's are R-ordered in some fashion, so we may assume that $s_0 R s_1 \cdots R s_m$. Then if $s_m Rt$, we cannot have $t \in x_j$ for any j, or else as $s_j R s_m$, we get $s_j Rt$, contrary to the definition of s_j. Thus

$$s_m Rt \quad \text{implies} \quad |t| \notin X.$$

Next observe that $\mathcal{M} \not\models_{s_m} \Box B$. For, since $\mathcal{M} \not\models_s \Box B$, $\mathcal{M} \not\models_u B$ for some $u \in S$ with sRu. Then $|s|R^\tau|u|$. But $|s_m|R^\tau|s|$, since $|s_m|$ and $|s|$ have the same R^τ-cluster C, so $|s_m|R^\tau|u|$, ensuring that $\mathcal{M} \not\models_{s_m} \Box B$ by filtration condition (F2).

Hence $\mathcal{M} \not\models_t B$ for some t such that $s_m Rt$. But then $|t| \notin X$, as above, so the R^c-cluster of $|t|$ does not precede that of z, implying that $zR^c|t|$, and hence $|s|R^c|t|$.

Finally, if $|s| \notin C^H$, then not $zR^c|s|$, and so not $|t|R^c|s|$.

Corollary 8.12. *If $\mathcal{M}^c = (S_\Gamma, R^c, V_\Gamma)$, then for any $B \in \Gamma$ and $s \in S$,*

$$\mathcal{M} \models_s B \quad \text{iff} \quad \mathcal{M}^c \models_{|s|} B.$$

Proof. Exercise.

Completeness for $S4.3Dum$

The construction of $\mathcal{M}^c$ just described will give a finite falsifying model for any non-theorem of any normal logic Λ containing $S4.3$. But to use this in a completeness theorem for Λ we would need to show that $\mathcal{M}^c$ validates Λ. In the case of $S4.3Dum$ we achieve this by using Dum to show that every non-last R^c-cluster is simple. For this purpose, a further general result is needed.

Definability Lemma 8.13. *For any $X \subseteq S_\Gamma$, there is a formula B_X(a truth-functional combination of members of Γ) such that for all $s \in S$,*

$$\mathcal{M} \models_s B_X \quad \text{iff} \quad |s| \in X.$$

Proof. For each $t \in S$, let B_t be the conjunction of the (finitely many) formulae in the set

$$\{B \in \Gamma : \mathcal{M} \models_t B\} \cup \{\neg B : B \in \Gamma \;\&\; \mathcal{M} \not\models_t B\}.$$

Then

$$\mathcal{M} \models_s B_t \quad \text{iff} \quad s \sim_\Gamma t \quad \text{iff} \quad |s| = |t|.$$

Now S_Γ is finite, since Γ is finite. So if

$$X = \{|t_1|, \ldots, |t_n|\},$$

we can take B_X to be

$$B_{t_1} \vee \cdots \vee B_{t_n}.$$

Dum-**Lemma 8.14.** *If $\mathcal{M} \models Dum$, then every non-last R^c-cluster is simple.*

Proof. Let C be a non-last R^c-cluster, and take $x \in C$. Then there must be some $y \in S_\Gamma$ such that xR^cy but not yR^cx.

By 8.13, there is a formula B that defines in $\mathcal{M}$ the set $X = \{s : |s| \neq x\}$. In other words, for all $s \in S$,

$$\mathcal{M} \models_s B \quad \text{iff} \quad s \notin x. \tag{i}$$

Since not yR^cx, there exists some $t \in y$ such that if tRu then $u \notin x$ and so $\mathcal{M} \models_u B$ by (i). Thus

$$\mathcal{M} \models_t \Box B. \tag{ii}$$

Now pick any $s \in x$. Then not tRs, so by R-connectedness sRt. Hence from (ii),

$$\mathcal{M} \models_s \Diamond\Box B. \tag{iii}$$

But $\mathcal{M} \not\models_s B$ by (i), so from (iii) and $\mathcal{M} \models Dum$ it follows that

$$\mathcal{M} \not\models_s \Box(\Box(B \to \Box B) \to B).$$

Hence there exists $u \in S$ with sRu,

$$\mathcal{M} \models_u \Box(B \to \Box B), \tag{iv}$$

and $\mathcal{M} \not\models_u B$. Whence $u \in x$ by (i).

Now suppose, for the sake of contradiction, that C is not simple. Then there exists some $z \in C$ with $z \neq x$. Thus xR^cz, and so as $u \in x$, uRv for some $v \in z$. It follows from (iv) that $\mathcal{M} \models_v B \to \Box B$. But $\mathcal{M} \models_v B$, since $v \in z \neq x$, so this leads to $\mathcal{M} \models_v \Box B$. However, since $z, x \in C$ we have zR^cx, so vRw for some $w \in x$. Then $\mathcal{M} \models_w B$, which is our desired contradiction in view of (i).

Finite Frame Property for $S4.3Dum$

By 8.12 and 8.14, each non-theorem of $S4.3Dum$ can be invalidated by a finite reflexive transitive and connected frame in which every non-last cluster is simple (has only one element). But every such frame is a p-morphic image of $(\omega, \leq)$ (by a construction similar to that in Theorem 8.4), and hence is an $S4.3Dum$-frame. This establishes the finite frame property for $S4.3Dum$, and the fact that the logic is determined by $(\omega, \leq)$. (At this point the reader could proceed directly to §9.)

Exercises 8.15

(1) Fill in all the details of the argument just given.

(2) A variant of Dum is the schema

$$Dum^* : \qquad \Box(\Box(A \to \Box A) \to A) \to (\Diamond\Box A \to \Box A).$$

Use the completeness theorem just given to show that $S4.3Dum = S4.3Dum^*$.

(3) Show that the smallest normal logic containing $S4.3$ and the schema

$$\Box(\Box(A \to \Box A) \to A) \to A$$

is determined by the class of finite reflexive total orderings, and also by the frame $(\omega, \geq)$. Show further that an alternative to this schema for the logic in question is

$$\Box(\Box(A \to \Box A) \to A) \to \Box A.$$

Bull's Theorem

One of the more striking results in the metatheory of propositional modal logic is that *every uniform normal extension of $S4.3$ has the finite model property*. This was shown by Bull [1966], using algebraic models (Boolean algebras with a unary operator interpreting $\Box$). A frame-theoretic argument was given by Fine [1971]. By utilising our description of the relationship between R^c and R^τ clusters, it is possible to give a very clear account of how Fine's proof works.

Let Λ be any uniform normal logic containing $S4.3$, i.e. containing the schemata T, 4, and L. Suppose that $\nvdash_\Lambda A$. We want then to construct a finite Λ-model that falsifies A. Let $\mathcal{M} = (S, R, V)$ be the submodel of the canonical Λ-model generated by some point s_A with $A \notin s_A$. Put $\Gamma = Sf(\Box A)$ (the reason for including $\Box A$ in Γ will be revealed later).

Now we saw in the completeness proof for $S4.3Dum$ that the models $\mathcal{M}^\tau$ and $\mathcal{M}^c$ will falsify A, but neither can be guaranteed to be a Λ-model.

To construct a Λ-model that rejects A, we will remove all but the head from each R^τ-cluster.

A point $x \in S_\Gamma$ is called *essential* if it belongs to the head of its own R^τ-cluster. Let

$$E = \{x : x \text{ is essential}\}$$

be the union of all the heads of R^τ-clusters. Notice that the relations R^τ and R^c are identical when restricted to E.

Now define a map $f : S \to E$, as follows. For each R^τ-cluster C, let x_C be a fixed, but arbitrarily chosen, member of the *head* of C. Put $f(s) = |s|$ if $|s|$ is essential, and otherwise let $f(s) = x_C$, where C is the R^τ-cluster of $|s|$. In both cases, $|s|$ and $f(s)$ are in the same R^τ-cluster, so $f(s)R^\tau|s|$ and $|s|R^\tau f(s)$. Moreover, since $f(s)$ is in the head, we invariably have $|s|R^c f(s)$.

Lemma 8.16. *f is a p-morphism from (S, R) onto (E, R^τ).*

Proof. First, if sRt, then $f(s)R^\tau|s|R^\tau|t|R^\tau f(t)$, and so $f(s)R^\tau f(t)$ as R^τ is transitive.

Second, suppose $f(s)R^\tau x$, with $x \in E$. Then $f(s)R^c x$. But $|s|R^c f(s)$, so $|s|R^c x$, implying that there is a t with sRt and $t \in x$, hence $f(t) = x$.

This establishes the two p-morphism conditions for f.

Lemma 8.17. *For any $Y \subseteq E$, there is a formula B_Y such that for all $s \in S$,*

$$\mathcal{M} \models_s B_Y \quad \text{iff} \quad f(s) \in Y.$$

Proof. By the way f was constructed, using the fixed elements x_C, f preserves Γ-equivalence classes, i.e.

$$t \in |s| \quad \text{implies} \quad f(t) = f(s).$$

Thus the set $\{s : f(s) \in Y\}$ is a union of equivalence classes, and so there is an $X \subseteq S_\Gamma$ such that

$$f(s) \in Y \quad \text{iff} \quad |s| \in X.$$

But then taking B_Y to be the formula B_X of Definability Lemma 8.13 gives the desired result.

The Finite Λ-Model

A model $\mathcal{M}_E = (E, R^\tau, V_E)$ on E is now defined by putting $V_E(p) = E$ if $p \notin \Gamma$, and otherwise

$$V_E(p) = \{|s| \in E : s \in V(p)\} = V_\Gamma(p) \cap E,$$

so that V_E is defined on the whole of $Fma(\Phi)$ (the definition on $\Phi - \Gamma$ is immaterial). By Lemma 8.17, for each $p \in \Phi$ there is a formula B_p such that

$$\mathcal{M} \models_s B_p \quad \text{iff} \quad f(s) \in V_E(p).$$

For any formula B, let B' be the result of uniformly replacing each atomic p occurring in B by B_p. Precisely:

$$p' = B_p, \quad \bot' = \bot, \quad (B \to D)' = B' \to D', \quad (\Box B)' = \Box(B').$$

Then for all $B \in Fma(\Phi)$, we get

$$\mathcal{M} \models_s B' \quad \text{iff} \quad \mathcal{M}_E \models_{f(s)} B. \tag{†}$$

The case $B = p \in \Phi$ of this result is given by the definition of B_p, and the inductive cases are straightforward, as f is a p-morphism.

It now follows that $\mathcal{M}_E$ is a Λ-model. For if $\vdash_\Lambda B$, then since Λ is a *uniform* logic, $\vdash_\Lambda B'$. Thus $\mathcal{M} \models B'$ (because $\mathcal{M}$ is a generated submodel of $\mathcal{M}^\Lambda$), and hence $\mathcal{M}_E \models B$ by (†). It remains to show that $\mathcal{M}_E$ rejects the non-Λ-theorem A.

Lemma 8.18. *If $\Box B \in \Gamma$ and $\mathcal{M} \not\models_s \Box B$, then $\mathcal{M} \not\models_t B$ for some t such that $|s| R^\tau |t|$ and $|t| \in E$.*

Proof. If $\mathcal{M} \not\models_s \Box B$, then $\mathcal{M} \not\models_t B$ for some t with sRt and so $|s|R^\tau|t|$. If $|t| \in E$, we are done. Otherwise, since $\mathcal{M} \models (\Box B \to B)$, we have $\mathcal{M} \not\models_t \Box B$, and so by the R^c-Lemma 8.11, $\mathcal{M} \not\models_u B$ for some u with $|u|$ in a later R^c-cluster than $|t|$. If $|u| \notin E$, we repeat the argument to obtain $\mathcal{M} \not\models_v B$ for some v with $|v|$ in a later R^c-cluster than $|u|$. Since the sequence of R^c-clusters is finite in length, this process cannot move forward ad infinitum, and must end, in the very last R^c-cluster if not before, with the desired conclusion.

Corollary 8.19. *If $|s| \in E$, then for any $B \in \Gamma$,*

$$\mathcal{M} \models_s B \quad \text{iff} \quad \mathcal{M}_E \models_{|s|} B.$$

Proof. The atomic case holds by definition of V_E. The inductive case of $\Box$ is taken care of by the definition of R^τ and Lemma 8.18.

It is now apparent why we put the formula $\Box A$ into Γ. For, since $\mathcal{M}$ is a T-model and $\mathcal{M} \not\models_{s_A} A$ we get $\mathcal{M} \not\models_{s_A} \Box A$, so by Lemma 8.18 there exists t such that $|t| \in E$ and $\mathcal{M} \not\models_t A$. Then by Corollary 8.19, $\mathcal{M}_E \not\models_{|t|} A$, showing that the finite Λ-model $\mathcal{M}_E$ falsifies the non-Λ-theorem A. This completes the proof of Bull's Theorem that every uniform normal logic containing $S4.3$ has the finite model property, and hence has the finite frame property (Exercise 4.9(5)).

Linear Temporal Logics

For the remainder of this section we return to the [F]-[P]-language of temporal logic, in which the formula $\Box A$ is introduced as short-hand for

$$[P]A \wedge A \wedge [F]A$$

(at this point it would be appropriate to review the material of §6).

A *linear temporal logic* is any normal logic in this language that contains the smallest temporal logic K_t, and also the schemata

$$\Box A \to [P][F]A,$$

and

$$\Box A \to [F][P]A.$$

The smallest linear temporal logic will be denoted *Lin*. In view of Exercises 6.3, it follows that *Lin* is determined by the class of transitive, weakly future-connected, and weakly past-connected frames. Indeed, the canonical model $\mathcal{M}^{\Lambda}$ of any linear temporal logic Λ has these properties. Hence any generated submodel $\mathcal{M}$ of such a canonical model is transitive and connected (Exercise 6.5(3)). Consequently, a temporal filtration $\mathcal{M}^{\tau}$ of such a generated subframe will also be transitive and connected ($\mathcal{M}^{\tau}$ was defined just prior to Exercises 6.6).

We will consider the completeness problem for the three standard types of irreflexive linear time.

Discrete Time

Let *LinDisc* be the smallest linear temporal logic containing the schemata

$$\begin{array}{ll} D_{\mathrm{F}}: & <\mathrm{F}> \top \\ D_{\mathrm{P}}: & <\mathrm{P}> \top \\ Z_{\mathrm{F}}: & [\mathrm{F}]([\mathrm{F}]A \to A) \to (<\mathrm{F}> [\mathrm{F}]A \to [\mathrm{F}]A) \\ Z_{\mathrm{P}}: & [\mathrm{P}]([\mathrm{P}]A \to A) \to (<\mathrm{P}> [\mathrm{P}]A \to [\mathrm{P}]A) \end{array}$$

Then *LinDisc* is determined by the integer frame $(\mathbb{Z}, <)$. The proof of this is a straightforward adaptation of the proof that the modal logic $K4DLZ$ is determined by the frame $(\omega, <)$. In *LinDisc* there is complete symmetry between the past and future operators. D_{F} makes the last cluster in a finite filtration $\mathcal{M}^{\tau}$ be non-degenerate, while D_{P} does the same to the first cluster. Z_{F} allows all non-last clusters to be modifiable without affecting the truth-values of formulae of the type $[\mathrm{F}]B$ from Γ. Similarly, by the Z_{P}-analogue of the Z-Lemma 8.5, Z_{P} allows all *non-first* clusters to be modified without affecting $[\mathrm{P}]B$-type formulae. So, we replace each cluster except the first and last by a strict total ordering of its elements, treat the last cluster in the same manner as in Theorem 8.4, and apply the mirror image of this treatment to the first cluster, to get a temporal p-morphism from $(\mathbb{Z}, <)$ onto the frame of $\mathcal{M}^{\tau}$.

Beginning Time

Let *LinDisc*$^{\omega}$ be the logic that results when the schema Z_{P} in the definition of *LinDisc* is replaced by

$$W_{\mathrm{P}}: \quad [\mathrm{P}]([\mathrm{P}]A \to A) \to [\mathrm{P}]A,$$

and D_{P} is deleted.

W_{P} allows *any* cluster in $\mathcal{M}^{\tau}$ to be modified without affecting truth of $[\mathrm{P}]B$-type formulae from Γ. Hence the $K4DLZ$ construction of Theorem 8.4 applies directly to show that *LinDisc*$^{\omega}$ is determined by the time-frame $(\omega, <)$. In fact, by including the formula $[\mathrm{P}]\bot$ in Γ, we can obtain this result using only the special case $A = \bot$ of W_{P}, for then the first cluster is already in the desired form:

Exercise 8.20

Let $\mathcal{M}^{\tau}$ be a finite temporal Γ-filtration of a generated submodel of the canonical *LinDisc*$^{\omega}$-model. Suppose $[\mathrm{P}]\bot \in \Gamma$. Then if $|s|$ belongs to the first cluster of $\mathcal{M}^{\tau}$, show that $[\mathrm{P}]\bot \in s$. Deduce that this first cluster is degenerate. (It might be useful here to note that when $A = \bot$, W_{P} is equivalent to

$$[\mathrm{P}]\bot \vee \langle\mathrm{P}\rangle[\mathrm{P}]\bot).$$

Rational Time

Let *LinRat* be the smallest normal extension of *Lin* that contains the schemata D_{F}, D_{P}, and

$$X_{\mathrm{F}}: \qquad [\mathrm{F}][\mathrm{F}]A \to [\mathrm{F}]A.$$

Then *LinRat* is the temporal logic determined by the rational-number frame $(\mathbb{Q}, <)$.

The effect of axioms D_{P} and D_{F} has already been noted: they force the first and last clusters to be non-degenerate in any finite filtration $\mathcal{M}^{\tau}$ of a generated submodel of the canonical *LinRat*-model. The effect of X_{F} is then to force any degenerate cluster in $\mathcal{M}^{\tau}$ to be immediately followed by a non-degenerate one (cf. Exercise 8.8(2)). Knowing this, the following result can be obtained, and gives the asserted completeness theorem.

Theorem 8.21. *There is a temporal p-morphism from* $(\mathbb{Q}, <)$ *onto the frame of* $\mathcal{M}^{\tau}$.

Proof. By a *rational open interval* we mean a set of the form

$$(r, q) = \{x \in \mathbb{Q} : r < x < q\},$$

where the end-point r is either $-\infty$ or a real number, and q is either $+\infty$ or a real (it is crucial to the construction that r and q need not be rational). Any rational open interval can be mapped by a temporal p-morphism onto any non-degenerate cluster in $\mathcal{M}^\tau$. For, since $(\mathbb{Q}, <)$ is dense, and serial in both directions, we can choose a subset $\{x_m : m \in \mathbb{Z}\}$ of (r, q) that is order-isomorphic to $(\mathbb{Z}, <)$, i.e. has

$$x_m < x_k \quad \text{iff} \quad m < k,$$

with the x_m's converging to q for increasingly positive m, and to r for increasingly negative m. Then if a non-degenerate cluster C has elements $z_1, \ldots, z_n$, a suitable temporal p-morphism f from $((r, q), <)$ onto C is given by putting

$$f(x_m) = z_j \quad \text{iff} \quad m \equiv j (\text{mod } n),$$

and otherwise letting $f(x)$ be any member of C.

Now let $C_1, \ldots, C_k$ be all the clusters of $\mathcal{M}^\tau$ in their $<$-order, as induced by R^τ. Since, by D_{P}, the first cluster C_1 is non-degenerate, we can map a rational open interval of the form $(-\infty, r_1)$ p-morphically onto C_1. Now if C_2 is non-degenerate, we take r_1 here to be an *irrational* real number, and then map a rational open interval of the form (r_1, r_2) onto C_2. If, on the other hand, C_2 is degenerate, we take r_1 to be rational, and map r_1 to the unique element of C_2. Since C_3 must now be non-degenerate, by axiom X_{F}, we can then map an interval of the form (r_1, r_2) p-morphically onto C_3. In either case, the interval $(-\infty, r_2)$ gets mapped into $\mathcal{M}^\tau$ without "missing out" any rationals less than r_2.

Continuing this process, we eventually come to the last cluster C_k, having mapped an interval of the form $(-\infty, r']$ or $(-\infty, r')$ p-morphically onto $C_1 \cup \cdots \cup C_{k-1}$, with the real number r' being irrational in the second case. But C_k is non-degenerate, by D_{F}, so the construction can be completed by mapping $(r', +\infty)$ p-morphically onto C_k.

Exercises 8.22

(1) Show that *LinRat* is determined by any rational open interval frame $((r, q) <)$.

(2) Give a proof-theoretic deduction of the schema

$$[\mathrm{P}][\mathrm{P}]A \to [\mathrm{P}]A$$

in *LinRat*.

(3) Show that the time-frame $(\mathbb{Q}, \leq)$ determines the smallest normal extension of *Lin* containing the schema

$$[\mathrm{F}]A \to A.$$

Real Time

LinRe is the smallest normal extension of *LinRat* that contains the schema

$$Cont: \qquad \Box([\mathrm{P}]A \to <\mathrm{F}> [\mathrm{P}]A) \to ([\mathrm{P}]A \to [\mathrm{F}]A).$$

Exercise 6.4(4) asked the reader to show that *Cont* is valid in the real-number time-frame $(\mathbb{R}, <)$. To prove that *LinRe* is determined by this frame, we adapt the argument given in Theorem 8.21 for *LinRat*, this time using *real* open intervals

$$(r, q) = \{x \in \mathbb{R} : r < x < q\},$$

where again the end-points are either reals, or $\pm\infty$.

Working now with a finite temporal Γ-filtration $\mathcal{M}^\tau$ of a generated submodel $\mathcal{M} = (S, R, V)$ of the canonical *LinRe*-model, we try to map $(\mathbb{R}, <)$ onto the frame of $\mathcal{M}^\tau$ by a temporal p-morphism. A problem comes up if we strike a non-degenerate cluster C in $\mathcal{M}^\tau$ that is immediately succeeded by a cluster D that is also non-degenerate. Having mapped a real open interval (r, q) p-morphically onto C, we cannot then treat D similarly without leaving out the end-point q.

This problem would not arise if in $\mathcal{M}^\tau$ there were no adjacent non-degenerate clusters (for D would then have to be degenerate, and we could map q to its unique element). However it does not seem possible to prevent pairs

$$\cdots \to \bigcirc \to \bigcirc \to \cdots$$

of adjacent non-degenerate clusters from occurring. Instead we will have to show that the model $\mathcal{M}^\tau$ has a certain property that allows it to be modified, by inserting a new degenerate cluster between any such pair, creating the configuration

$$\cdots \to \bigcirc \to \bullet \to \bigcirc \to \cdots$$

and thereby removing the problem – without altering the truth-values of members of Γ at any of the old points of $\mathcal{M}^\tau$. The idea of this construction comes from Segerberg [1970], although the axiom *Cont* we use, and the argument in which it is applied (in Lemma 8.23), are different.

So, let C and D be non-degenerate clusters in $\mathcal{M}^\tau$ that are adjacent, with $C < D$. An element s of the sub-canonical model $\mathcal{M}$ will be called *C-greatest* if

$$|s| \in C, \text{ and } \forall t \in S\,(sRt \text{ implies } |t| \notin C).$$

Dually, s is *D-least* if

$$|s| \in D, \text{ and } \forall t \in S\,(tRs \text{ implies } |t| \notin D).$$

These notion may be intuitively related to the situation in the real-number frame $(\mathbb{R}, <)$, where the element z that fills a cut (X, Y) (i.e. has $x \leq z \leq y$ for $x \in X$ and $y \in Y$) must be either a greatest element of X, or a least element of Y.

Lemma 8.23. *There exists an element of $\mathcal{M}$ that is either C-greatest or D-least.*

Proof. Suppose that there is no C-greatest element, and no D-least element. Let A be a formula such that for all $s \in S$,

$$\mathcal{M} \models_s A \quad \text{iff} \quad C_{|s|} \leq C,$$

where $C_{|s|}$, as usual, is the R^{τ}-cluster of $|s|$. Such a formula exists by the argument of the Definability Lemma 8.13.

Sublemma A. $\mathcal{M} \models_s [\mathrm{P}]A$ *iff* $C_{|s|} \leq C$.

Proof. Suppose $C_{|s|} \leq C$. Then if tRs, $|t|R^{\tau}|s|$, so $C_{|t|} \leq C_{|s|} \leq C$, and hence $\mathcal{M} \models_t A$ by definition of A. This shows $M \models_s [\mathrm{P}]A$.

For the converse, suppose $C_{|s|} \not\leq C$. Then by $<$-connectedness, $C < C_{|s|}$, and so as D is adjacent to C, $D \leq C_{|s|}$. Take first the case that $D = C_{|s|}$, so that $|s| \in D$. Since, by assumption, s is not D-least, there must exist a t with tRs and $|t| \in D$. But D comes after C, so $\mathcal{M} \not\models_t A$, by definition of A. Hence $\mathcal{M} \not\models_s [\mathrm{P}]A$. On the other hand, if $D < C_{|s|}$, then taking any t with $|t| \in D$ must give tRs by R-connectedness, and so the same argument applies to give $\mathcal{M} \not\models_s [\mathrm{P}]A$.

Sublemma B. $\mathcal{M} \models \Box([\mathrm{P}]A \to \langle \mathrm{F} \rangle [\mathrm{P}]A)$.

Proof. It suffices to prove

$$\mathcal{M} \models [\mathrm{P}]A \to \langle \mathrm{F} \rangle [\mathrm{P}]A.$$

So, suppose $\mathcal{M} \models_s [\mathrm{P}]A$. Then by Sublemma A, $C_{|s|} \leq C$. If $C_{|s|} = C$, then since, by assumption, s is not C-greatest, there exists t with sRt and $|t| \in C$. But then Sublemma A again gives $\mathcal{M} \models_t [\mathrm{P}]A$, and so $\mathcal{M} \models_s \langle \mathrm{F} \rangle [\mathrm{P}]A$. On the other hand, if $C_{|s|} < C$, then any t with $|t| \in C$ will have sRt, and the same conclusion follows.

Finally, to complete the proof of Lemma 8.23, take any s with $|s| \in C$. Then $\mathcal{M} \models_s [\mathrm{P}]A$. Choose any t with $|t| \in D$. As in the proofs of the Sublemmas, we then get sRt and $\mathcal{M} \not\models_t A$, so $\mathcal{M} \not\models_s [\mathrm{F}]A$. Hence

$$\mathcal{M} \not\models_s [\mathrm{P}]A \to [\mathrm{F}]A.$$

But in view of Sublemma B, this contradicts the fact that $\mathcal{M} \models Cont$.

Filling The Cut

Armed now with Lemma 8.23, we enlarge the model $\mathcal{M}^\tau$ to a new model $\mathcal{M}'$ as follows. Suppose that there is a C-greatest element γ in $\mathcal{M}$. $\mathcal{M}'$ is then based on the frame (S', R'), where

$$S' = S_\Gamma \cup \{\gamma\}, \qquad \text{and}$$
$$R' = R^\tau \cup \{(x,\gamma) : x \in S_\Gamma \ \&\ C_x \le C\} \cup \{(\gamma, x) : x \in S_\Gamma \ \&\ D \le C_x\}.$$

Let $f : S' \to S_\Gamma$ have $f(\gamma) = |\gamma|$, and otherwise $f(x) = x$. For each $x \in S'$, put

$$\mathcal{M}' \models_x p \quad \text{iff} \quad \mathcal{M}^\tau \models_{f(x)} p.$$

Thus $\mathcal{M}'$ arises by inserting γ as a new irreflexive element (degenerate cluster) between C and D in $\mathcal{M}^\tau$.

Lemma 8.24. *For any $B \in \Gamma$, and any $x \in S'$,*

$$\mathcal{M}' \models_x B \quad \text{iff} \quad \mathcal{M}^\tau \models_{f(x)} B.$$

Proof. Since C is non-degenerate, we have that

$$xR'y \quad \text{implies} \quad f(x)R^\tau f(y), \tag{$\dagger$}$$

for all $x, y \in S'$, and this suffices to prove the inductive cases of $[\mathrm{P}]B$ and $[\mathrm{F}]B$ from right to left. For the converses, suppose first that $\mathcal{M}^\tau \not\models_{f(x)} [\mathrm{F}]B$. A little reflection reveals that the only problematic case is when $x = \gamma$. But then $f(x) = |\gamma|$, and so by the Filtration Lemma, $\mathcal{M} \not\models_\gamma [\mathrm{F}]B$, hence $\mathcal{M} \not\models_t B$ for some t with γRt. Since γ is C-greatest, $|t| \notin C$, so as $|\gamma|R^\tau|t|$, $C < C_{|t|}$, and therefore $\gamma R'|t|$ in $\mathcal{M}'$. But $\mathcal{M}^\tau \not\models_{|t|} B$, so applying the induction hypothesis on B, we then get $\mathcal{M}' \not\models_\gamma [\mathrm{F}]B$.

Finally the inductive case that $\mathcal{M}' \models_x [\mathrm{P}]B$ implies $\mathcal{M}^\tau \models_{f(x)} [\mathrm{P}]B$ is straightforward, since ($\dagger$) holds for all $x \in S'$ when $y = \gamma$.

Exercise 8.25

Adapt the construction of $\mathcal{M}'$ to the case that there is instead a D-least element, and prove Lemma 8.24 for that case.

Now by iterating the $\mathcal{M}'$-construction a finite number of times, we obtain a model $\mathcal{M}''$ with no adjacent non-degenerate clusters, no adjacent degenerate ones, and non-degenerate first and last clusters. A temporal p-morphism can then be constructed from $(\mathbb{R}, <)$ onto the frame of $\mathcal{M}''$, as discussed above. But by Lemma 8.24 (iterated), any Γ-formula falsifiable in $\mathcal{M}^\tau$ is falsifiable in $\mathcal{M}''$, and hence falsifiable in a model based on $(\mathbb{R}, <)$.

Exercise 8.26

Axiomatise the temporal logics determined by $(\mathbb{Z}, \le)$, $(\omega, \le)$, and $(\mathbb{R}, \le)$ (cf. Segerberg [1970] for some answers).

9 | Temporal Logic of Concurrency

Consider the following description of a "concurrent" program (adapted from Pnueli [1981]). There are n different processes acting in parallel, using a shared memory environment, so that each can alter the values of variables used by the others. For illustrative purposes, the processes may be thought of as disjoint flowcharts, with labelled nodes. A typical node of the i-th process is denoted m^i. Each process has an entry node m_0^i. If the program variables are $v_1, \ldots, v_k$, then a *state* may be defined as a vector

$$s = (m^1, \ldots, m^n, a_1, \ldots, a_k),$$

specifying a label for each process (denoting the point that the process is currently at), and a current value a_i for each variable v_i. Predicates at_i of labels will be used, with the semantics

$$\models_s at_i(m) \quad \text{iff} \quad m = m^i.$$

Each successive state is to be obtained from its predecessor by exactly one process being chosen to execute one transition in its flow chart. Thus from an initial state

$$s_0 = (m_0^1, \ldots, m_0^n, a_1, \ldots, a_k),$$

many different execution sequences $s_0, s_1, \ldots \ldots$ may be generated, depending on which process gets chosen to act at each step. Some interesting properties of such sequences can be formulated by reading the connective $\Box$ as "at all states from now on".

Deadlock Freedom

Deadlock occurs when no processor can act. The requirement that deadlock does not occur at $(m^1, \ldots, m^n)$ can be expressed by

$$\Box(at_1(m^1) \wedge \cdots \wedge at_n(m^n) \to E_1 \vee \cdots \vee E_n),$$

where E_i is the *exit condition* for node m^i consisting of the disjunction of the propositions labelling edges out of m^i (the truth of such a proposition being the requirement for the process to be able to proceed along that edge).

Mutual Exclusion

$$\Box\neg(at_i(m) \land at_j(m'))$$

asserts that the program can never simultaneously access m and m'.

Accessibility

$$\Box(at_i(m) \to \Diamond at_j(m'))$$

expresses that if the program ever reaches m it will eventually proceed from there to m'.

Correctness

A *partial correctness* assertion about a program states that if the program works as was intended, then a certain condition ψ must be true after termination, given that some condition φ was true at the start. Illustrating with a program having a single entry label m_0, and exit m_e, this can be formalised as

$$at(m_0) \land \varphi \to \Box(at(m_e) \to \psi).$$

Total correctness includes the assertion that the program will halt:

$$at(m_0) \land \varphi \to \Diamond(at(m_e) \land \psi).$$

Responsiveness

An operating system may receive requests (r_i) from various agents, to whom it will signal (q_i) when it grants the request. The formula

$$\Box(r_i \to \Diamond q_i)$$

expresses that a request is always eventually honoured.

Absence of Unsolicited Response

This example, from Gabbay et. al. [1980], uses the connective $\mathcal{U}$ (until) to express the requirement that if a response is to occur, it will not do so until a request has been received:

$$\Diamond q_i \to (\neg q_i)\mathcal{U}r_i.$$

Further explanations of how temporal logic is used in applications to computer science may be found in Manna and Pnueli [1981], Hailpern [1982], Moszkowski [1986], and several articles in Galton [1987] and de Bakker et. al. [1989].

Syntax and Semantics

Given a set Φ of atomic formulae as usual, define a set of formulae $A \in Fma(\Phi)$ by the BNF definition

$$A ::= p \mid \bot \mid A_1 \to A_2 \mid \Box A \mid \bigcirc A \mid A_1 \mathcal{U} A_2$$

$\Box$ means "henceforth" (i.e. from now on, including the present).
$\bigcirc$ means "next" (i.e. at the next state).
$\mathcal{U}$ means "until".
$\Diamond$, as usual, is shorthand for $\neg\Box\neg$.

By a *state sequence* we mean a pair $\mathcal{F} = (S, \sigma)$, where σ is a surjective function $\omega \to S$ enumerating S as a sequence

$$\sigma_0, \sigma_1, \ldots, \sigma_n, \ldots\ldots$$

(possibly with repetition, for example when S is finite).

A *model* $\mathcal{M} = (S, \sigma, V)$ on a state sequence is defined in the usual way, and the relation

$$\mathcal{M} \models_j A,$$

meaning "A is true at the j-th state σ_j in $\mathcal{M}$ ", is defined by

$\mathcal{M} \models_j p$	iff	$\sigma_j \in V(p)$
$\mathcal{M} \not\models_j \bot$		
$\mathcal{M} \models_j A \to B$	iff	$\mathcal{M} \models_j A$ implies $\mathcal{M} \models_j B$
$\mathcal{M} \models_j \bigcirc A$	iff	$\mathcal{M} \models_{j+1} A$
$\mathcal{M} \models_j \Box A$	iff	for all $k \geq j$, $\mathcal{M} \models_k A$
$\mathcal{M} \models_j A\mathcal{U}B$	iff	for some $k \geq j$, $\mathcal{M} \models_k B$ and for every i such that $j \leq i < k$, $\mathcal{M} \models_i A$.

The definitions of the relations $\mathcal{M} \models A$ and $\mathcal{F} \models A$ are as usual.

Intuitively, this semantics amounts to interpreting $\Box$ by the relation $\leq$, and $\bigcirc$ by the relation R, where jRk iff $k = j + 1$. R is functional, and the connection between the two relations is that $\leq$ is the *ancestral* (reflexive transitive closure) of R (cf. §1 for the definition of the ancestral). This observation is the key to the completeness theorem to follow.

Axioms

Let Θ be the smallest logic in the language just described that contains the schemata

$$
\begin{array}{ll}
K: & \Box(A \to B) \to (\Box A \to \Box B) \\
K_{\circ}: & \bigcirc(A \to B) \to (\bigcirc A \to \bigcirc B) \\
Fun: & \bigcirc\neg A \leftrightarrow \neg\bigcirc A \\
Mix: & \Box A \to A \land \bigcirc\Box A \\
Ind: & \Box(A \to \bigcirc A) \to (A \to \Box A) \\
\mathcal{U}1: & A\mathcal{U}B \to \Diamond B \\
\mathcal{U}2: & A\mathcal{U}B \leftrightarrow B \lor (A \land \bigcirc(A\mathcal{U}B))
\end{array}
$$

and is closed under Necessitation for $\Box$ and $\bigcirc$, i.e.,

$$A \in \Theta \quad \text{implies} \quad \Box A, \bigcirc A \in \Theta.$$

The roles of K, $K_{\circ}$, and the Necessitation rules are now familiar. The axiom Fun expresses the interpretation of $\bigcirc$ by a total function, while Mix and Ind together correspond to the interpretation of $\Box$ by the reflexive transitive closure of the interpretation of $\bigcirc$. The reflexivity schema T : $\Box A \to A$ is immediately implied by Mix. For the transitivity schema 4, see Theorem 9.2 below. Ind by itself expresses the *induction* principle that any set which contains σ_j and is closed under the taking of successor states must contain all states from σ_j on.

Exercises 9.1

(1) (*Soundness*). Prove that $\mathcal{F} \models \Theta$ for any state sequence $\mathcal{F}$.

(2) Show that $\vdash_\Theta A \to \bigcirc A$ implies $\vdash_\Theta A \to \Box A$.

(3) $\vdash_\Theta \bigcirc\Box A \to \bigcirc A$.

(4) $\vdash_\Theta \Box A \to \bigcirc A$.

Theorem 9.2. *The following schemata are Θ-derivable.*

(1) 4: $\Box \to \Box\Box A$

(2) $\bigcirc\Box A \to \Box\bigcirc\Box A$

(3) $\bigcirc\Box A \to \bigcirc(A \land \bigcirc\Box A)$

(4) $A \land \bigcirc\Box A \to \Box A$

(5) $\bigcirc\Box A \to \Box(A \to \Box A)$

(6) Dum: $\Box(\Box(A \to \Box A) \to A) \to (\Diamond\Box A \to A)$

Proof. We indicate the main points. The rest involves tautological reasoning, and principles that hold for all normal logics.

(1) From $\Box A \to \bigcirc\Box A$ (by *Mix*) and Exercise 9.1(2).

(2) From *Mix*, K_o-principles, and 9.1(2).

(3) Use 9.1(3), *Mix*, and K_o-principles.

(4) Using (3) and 9.1(2) gives

$$\vdash_\Theta A \wedge \bigcirc\Box A \to \Box(A \wedge \bigcirc\Box A).$$

But $\vdash \Box(A \wedge \bigcirc\Box A) \to \Box A$.

(5) From (4),

$$\vdash_\Theta \bigcirc\Box A \to (A \to \Box A),$$

and hence

$$\vdash_\Theta \Box\bigcirc\Box A \to \Box(A \to \Box A).$$

Then use (2).

(6) An instance of schema K is

$$\Box(\Box(A \to \Box A) \to A) \to (\Box\Box(A \to \Box A) \to \Box A),$$

which by result (5) and the schema 4 yields

$$\vdash_\Theta \Box(\Box(A \to \Box A) \to A) \to (\bigcirc\Box A \to \Box A). \qquad (\dagger)$$

Now an instance of *Ind* is

$$\Box(\neg\Box A \to \bigcirc\neg\Box A) \to (\neg\Box A \to \Box\neg\Box A),$$

which, with the help of *Fun*, leads to

$$\vdash_\Theta \Box(\neg\Box A \to \neg\bigcirc\Box A) \to (\neg\Box A \to \neg\Diamond\Box A),$$

and hence

$$\vdash_\Theta \Box(\bigcirc\Box A \to \Box A) \to (\Diamond\Box A \to \Box A).$$

But this, together with ($\dagger$) and schemata 4 and T, yields $\vdash_\Theta$ *Dum*.

Deriving L_1

The schema *Dum* will be used in the completeness theorem for Θ, along with

$$L_1: \quad \Box(\Box A \to B) \vee \Box(\Box B \to A),$$

which is also Θ-derivable. The following exercises give a guided tour of a proof of this which is due to Martin Abadi.

Exercises 9.3

Let X be $(\Box A \to B)$ and Y be $(\Box B \to A)$. Define the following formulae.

$$
\begin{array}{ll}
L_1 : & \Box X \vee \Box Y \\
A_1 : & \Box X \vee X \vee \Box Y \vee \neg Y \vee \bigcirc\Box Y \\
A_2 : & \Box Y \vee Y \vee \Box X \vee \neg X \vee \bigcirc\Box X \\
A_3 : & \Box X \vee X \vee \Box Y \vee Y \\
A_4 : & \Box X \vee \neg X \vee \bigcirc\Box X \vee \Box Y \vee \neg Y \vee \bigcirc\Box Y
\end{array}
$$

(1) Show that A_1 and A_2 are deducible in any logic that is $\Box$-normal.

(2) Show that A_3 is deducible in any logic containing the schema T for $\Box$.

(3) Use Theorem 9.2(4) and tautological reasoning to show that

$$\vdash_\Theta A_1 \wedge A_2 \wedge A_3 \wedge A_4 \to L_1.$$

Conclude that

$$\vdash_\Theta A_4 \to L_1.$$

(4) Use *Fun* to Θ-deduce

$$\bigcirc L_1 \to A_4.$$

(5) With the help of the last two results, obtain

$$\vdash_\Theta \neg A_4 \to \bigcirc\neg A_4,$$

and then use *Ind* to get

$$\vdash_\Theta \neg L_1 \to \Box\neg A_4.$$

(6) Show that $\Diamond A_4$ is deducible in any $\Box$-normal logic containing schema T for $\Box$. Conclude that

$$\vdash_\Theta L_1.$$

To gain an intuitive understanding of the formulae $A_1, \ldots, A_4$ that collectively imply L_1, suppose that L_1 were false at some state. Then both conjuncts of

$$\Diamond\neg X \wedge \Diamond\neg Y$$

would be true. For each conjunct, the state at which X (respectively, Y) will be false could either be the present state, or some future state, in which case the conjunct is still true at the next state. This gives four possible situations, each of which falsifies one of $A_1, \ldots, A_4$.

Exercises 9.4

(1) Show that

$$\vdash_\Theta A \to B \wedge \bigcirc A$$

implies

$$\vdash_\Theta A \to \Box B.$$

(2) Show that

$$\vdash_\Theta A \to \Diamond B \wedge (B \vee (D \wedge \bigcirc A)),$$

implies

$$\vdash_\Theta A \to D\mathcal{U}B.$$

(3) The following are Θ-deducible:

$$\bigcirc A \to \Diamond A$$
$$\bigcirc(A \vee B) \to \bigcirc A \vee \bigcirc B$$
$$\Box\bigcirc A \leftrightarrow \bigcirc\Box A$$
$$\Diamond\bigcirc A \leftrightarrow \bigcirc\Diamond A$$
$$\Diamond A \leftrightarrow A \vee \bigcirc\Diamond A$$
$$(\neg A)\mathcal{U}A \leftrightarrow \Diamond A$$
$$\Box A \wedge \Diamond B \to A\mathcal{U}B$$

Induction Models

An *induction frame* is a structure $\mathcal{F} = (S, f)$, with $f : S \to S$, i.e. f is a function from S to S. The "graph"

$$\{(s,t) : t = f(s)\}$$

of f is denoted R_f. R_f^* is the *ancestral* of R_f (§1). Thus sR_f^*t iff there is an R_f-*list* linking s to t, i.e. a finite sequence $s = s_0, \ldots, s_n = t$, with $f(s_i) = s_{i+1}$ for all $i < n$. Models on induction frames give a semantics for Θ, as follows.

$\mathcal{M} \models_s \bigcirc A$	iff	$\mathcal{M} \models_{f(s)} A$
$\mathcal{M} \models_s \Box B$	iff	sR_f^*t implies $\mathcal{M} \models_t A$
$\mathcal{M} \models_s A\mathcal{U}B$	iff	there exists an R_f-list $s = s_0, \ldots, s_k$, with $\mathcal{M} \models_{s_k} B$, and $\mathcal{M} \models_{s_i} A$ whenever $0 \leq i < k$.

Exercise 9.5

If $\mathcal{M}$ is an induction model, show that $\mathcal{M} \models \Theta$.

Completeness of Θ.

Fix a formula A such that $\nvdash_\Theta A$. We want to find a falsifying model for A on a state sequence, and for this we adapt the canonical model construction.

The relations $R^\Theta_\Box$ and $R^\Theta_\circ$ on the set S^Θ of Θ-maximal subsets of $Fma(\Phi)$ are given by

$$\begin{array}{lll} sR^\Theta_\Box t & \text{iff} & \{B : \Box B \in s\} \subseteq t, \quad \text{and} \\ sR^\Theta_\circ t & \text{iff} & \{B : \bigcirc B \in s\} \subseteq t. \end{array}$$

Now $R^\Theta_\Box$ is reflexive (since Mix implies the schema T for $\Box$), transitive (since schema 4 is Θ-derivable), and weakly-connected (since schema L_1, and hence L, is Θ-derivable). By Fun, $R^\Theta_\circ$ is functional.

Since $\nvdash_\Theta A$, there is some $s_A \in S^\Theta$ with $A \notin s_A$. Let

$$S = \{u \in S^\Theta : s_A(R^\Theta_\Box)^* u\}.$$

As $\vdash_\Theta \Box B \to \bigcirc B$,

$$uR^\Theta_\circ v \quad \text{implies} \quad uR^\Theta_\Box v,$$

so S is closed under $R^\Theta_\circ$, i.e.

$$u \in S \ \& \ uR^\Theta_\circ v \quad \text{implies} \quad v \in S.$$

Also, when restricted to S, $R^\Theta_\Box$ is reflexive, transitive, and *connected* (cf. Exercise 3.11(1)).

We will work with the structure

$$\mathcal{F} = (S, R^\Theta_\Box, R^\Theta_\circ),$$

which is in essence the subframe of the canonical Θ-frame generated by s_A. But $R^\Theta_\Box$ is not the ancestral of $R^\Theta_\circ$ (cf. Exercise 9.6(2) below), and we will have to collapse $\mathcal{F}$ by filtration to achieve that property. Moreover, we cannot work with the canonical model on $\mathcal{F}$, since it is not apparent that the Truth-Lemma (3.3) can be proved for formulae involving the connective $\mathcal{U}$. Instead therefore, we work directly with the relation of membership of Θ-maximal sets, using such properties as

$$\begin{array}{lll} \Box B \in s & \text{iff} & \forall t \in S\,(sR^\Theta_\Box t \text{ implies } B \in t), \\ \bigcirc B \in s & \text{iff} & \forall t \in S\,(sR^\Theta_\circ t \text{ implies } B \in t), \\ \vdash_\Theta B & \text{implies} & B \in s, \quad \text{for all } s \in S \end{array}$$

(cf. Exercise 2.3(1), Theorem 3.2, etc.).

Exercises 9.6

(1) Show that $(R^{\Theta}_{\circ})^* \subseteq R^{\Theta}_{\Box}$.

(2) Show that the set

$$t_0 = \{\bigcirc^n p : n \geq 0\} \cup \{\neg\Box p\}$$

is Θ-consistent, by showing that each finite subset of t_0 is true at some point of some Θ-model. Deduce that there exist $t, u \in S^{\Theta}$ with $tR^{\Theta}_{\Box}u$ but not $t(R^{\Theta}_{\circ})^*u$.

Filtration

Our filtration set Γ will have to contain more than just the subformulae of A. We define

$$\begin{aligned}\Gamma =& Sf(A) \cup \{\bigcirc\Box B : \Box B \in Sf(A)\}\\ &\cup \{\bigcirc(B\mathcal{U}D), \Box\neg D, \bigcirc\Box\neg D, \neg D : B\mathcal{U}D \in Sf(A)\}.\end{aligned}$$

Then Γ is finite: it has fewer than $6n$ elements, where n is the number of elements of $Sf(A)$. The purpose of this definition is to ensure that Γ has the following closure properties:

$A \in \Gamma$;
Γ is closed under subformulae;
$\Box B \in \Gamma$ implies $\bigcirc\Box B \in \Gamma$;
$B\mathcal{U}D \in \Gamma$ implies $\bigcirc(B\mathcal{U}D), \Box\neg D \in \Gamma$.

The definition of Γ-filtration is adapted as follows.

$$\begin{aligned}&s \sim_{\Gamma} t \quad \text{iff} \quad s \cap \Gamma = t \cap \Gamma,\\ &|s| = \{t : s \sim_{\Gamma} t\},\\ &S_{\Gamma} = \{|s| : s \in S\}.\end{aligned}$$

Definability Lemma 9.7. *If $X \subseteq S_{\Gamma}$, there is a formula B_X such that for all $s \in S$,*

$$B_X \in s \quad \text{iff} \quad |s| \in X.$$

Proof. For each $t \in S$, let B_t be the conjunction of the members of

$$\{B \in \Gamma : B \in t\} \cup \{\neg B : B \in \Gamma \ \&\ B \notin t\},$$

and then if

$$X = \{|t_1|, \ldots, |t_n|\},$$

put

$$B_X = B_{t_1} \vee \cdots \vee B_{t_n}$$

(the construction is just as for the Definability Lemma 8.13).

Now a relation $R_\Box$ on S_Γ is defined to be a *Γ-filtration* of $R_\Box^\Theta$ if, and only if,

(F1) $sR_\Box^\Theta t$ implies $|s|R_\Box|t|$, and

(F2) $|s|R_\Box|t|$ implies $\{B : \Box B \in s \cap \Gamma\} \subseteq t$.

Replacing $\Box$ by $\bigcirc$ throughout this definition gives the notion of a Γ-filtration $R_\circ$ of $R_\circ^\Theta$.

Ancestral Lemma 9.8. *If a relation $R_\circ$ on S_Γ is a Γ-filtration of $R_\circ^\Theta$, then the ancestral $R_\circ^*$ of $R_\circ$ is a Γ-filtration of $R_\Box^\Theta$.*

Proof.

(F1). Let $s \in S$. To show that $sR_\Box^\Theta t$ implies $|s|R_\circ^*|t|$, let

$$X_s = \{x \in S_\Gamma : |s|R_\circ^* x\}.$$

First we prove

$$\Box(A_s \to \bigcirc A_s) \in s, \qquad (\dagger)$$

where A_s is a formula, given by the Definability Lemma 9.7., having

$$A_s \in u \quad \text{iff} \quad |s|R_\circ^*|u|.$$

To prove this, suppose that $sR_\Box^\Theta t$ and $A_s \in t$. We want $\bigcirc A_s \in t$ to conclude (†). But $|s|R_\circ^*|t|$, by the definition of A_s, so $|s|R_\circ^n|t|$ for some $n \geq 0$. Then if $tR_\circ^\Theta u$, we have $|t|R_\circ|u|$, since $R_\circ$ is a Γ-filtration of $R_\circ^\Theta$, and so $|s|R_\circ^{n+1}|u|$. This gives $|u| \in X_s$, and hence $A_s \in u$. We have thus shown

$$tR_\circ^\Theta u \quad \text{implies} \quad A_s \in u,$$

and hence $\bigcirc A_s \in t$ as required.

Since s contains all instances of the induction axiom *Ind*, (†) then yields

$$(A_s \to \Box A_s) \in s.$$

But $A_s \in s$, since $|s|R_\circ^*|s|$, and so

$$\Box A_s \in s.$$

Hence if $sR_\Box^\Theta t$, then $A_s \in t$, and so $|s|R_\circ^*|t|$.

(F2). We want to prove that

$$|s|R_o^*|t| \quad \text{implies} \quad \{B : \Box B \in s \cap \Gamma\} \subseteq t.$$

First we show, for all $n \geq 0$, that

$$|s|R_o^n|t| \quad \text{implies} \quad \{\Box B : \Box B \in s \cap \Gamma\} \subseteq t. \qquad (\ddagger)$$

The case $n = 0$ is immediate, since $|s| = |t|$ implies $s \cap \Gamma = t \cap \Gamma$. Assuming the result for n, suppose $|s|R_o^{n+1}|t|$. Then $|s|R_o^n|u|$ and $|u|R_o|t|$, for some u. Thus if $\Box B \in s \cap \Gamma$, we have $\Box B \in u$ by the hypothesis on n, and so $\bigcirc\Box B \in u \cap \Gamma$ by the axiom *Mix* and the definition of Γ. But then $\Box B \in t$, as R_o is a Γ-filtration of R_o^Θ. This completes the inductive proof of ($\ddagger$).

Finally then, if $|s|R_o^*|t|$, we have $|s|R_o^n|t|$ for some n, so that if $\Box B \in s \cap \Gamma$, ($\ddagger$) gives $\Box B \in t$, and then *Mix* gives $B \in t$.

This completes the proof of the Ancestral Lemma, a result which substantiates the earlier remark that the axioms *Mix* and *Ind* characterise the interpretation of $\Box$ as the ancestral of the interpretation of $\bigcirc$. This fact will feature again in the study of dynamic logic in the next section.

The Role of *Fun*

The axiom *Fun* ensures that R_o^Θ is a *functional* relation, but this property may be lost in passing to R_o. To deal with this, we will use the *smallest* Γ-filtration of R_o^Θ, defined by

$$|s|R_o|t| \quad \text{iff} \quad \exists s' \in |s| \, \exists t' \in |t| \, (s' R_o^\Theta t').$$

***Fun*-Lemma 9.9.** *Let R_o be the smallest Γ-filtration of R_o^Θ. Then if $\bigcirc B \in \Gamma$ and $s \in S$, the following are equivalent.*
(1) $\bigcirc B \in s$.
(2) $\forall t(|s|R_o|t|$ implies $B \in t)$.
(3) $\exists t(|s|R_o|t|$ and $B \in t)$.

Proof. First, note that (1) and (2) are equivalent for any filtration of R_o^Θ.

Next, *Fun* guarantees that R_o^Θ is serial, and hence R_o is serial. But this is enough to make (2) imply (3).

Finally, assume (3). Then there are $s' \in |s|$ and $t' \in |t|$ with $s' R_o^\Theta t'$. Thus if $\bigcirc B \notin s$, then $\bigcirc B \notin s'$, as $s \sim_\Gamma s'$, and so $\bigcirc\neg B \in s'$ by *Fun*. But then $\neg B \in t'$, contradicting the fact that $B \in t$ and $t \sim_\Gamma t'$. Hence (1) must hold.

Exercise 9.10

Show that the *Fun*-Lemma holds for any Γ-filtration of $R_{\circ}^{\Theta}$, smallest or not, provided that

if $\bigcirc B \in \Gamma$, then either $\bigcirc\neg B \in \Gamma$, or else $B = \neg C$ with $\bigcirc C \in \Gamma$.

Show that Γ can be made to satisfy this additional condition and still be finite.

Another way to explain the main point of the *Fun*-Lemma is that, under its hypotheses, if $\bigcirc B \notin s$, then $B \notin t$ for *any* t with $|s|R_{\circ}|t|$. The import of this will be that although $|s|$ may have a number of $R_{\circ}$-alternatives, we can remove all but one of them, in an arbitrary way, without altering the falsity of Γ-formulae of the form $\bigcirc B$ at $|s|$.

The Role of *Dum*

Consider the properties of the structure

$$(S_{\Gamma}, R_{\circ}, R_{\circ}^{*}),$$

where $R_{\circ}$ is the smallest Γ-filtration of $R_{\circ}^{\Theta}$. $R_{\circ}$ is serial, since $R_{\circ}^{\Theta}$ is, but may not be functional. $R_{\circ}^{*}$ is reflexive and transitive (by definition), and also *connected*, by the Ancestral Lemma, since $R_{\square}^{\Theta}$ is connected. Since S_{Γ} is finite, it follows that the structure takes the form of a finite sequence of $R_{\circ}^{*}$-clusters.

We now recall the analysis of extensions of $S4.3$ given in §8, and define the relation $R_{\circ}^{c}$ on S_{Γ} by

$$xR_{\circ}^{c}y \quad \text{iff} \quad \forall s \in x\, \exists t \in y\, (sR_{\square}^{\Theta}t).$$

Then $R_{\circ}^{c}$ is reflexive, transitive, and connected, with $R_{\circ}^{c} \subseteq R_{\circ}^{*}$ (this is proven just as in Lemma 8.10). Thus the $R_{\circ}^{c}$-cluster of each point is contained in the $R_{\circ}^{*}$-cluster of that point, and so each $R_{\circ}^{*}$-cluster decomposes into a sequence of $R_{\circ}^{c}$-clusters, as in the diagram on page 72. Moreover, the following result can be proved just as for the R^{c}-Lemma 8.11.

$R_{\circ}^{c}$-Lemma 9.11. *If $\square B \in \Gamma$ and $\square B \notin s \in S$, then there exists $t \in S$ with $B \notin t$ and $|s|R_{\circ}^{c}|t|$.*

Now by Theorem 9.2(6), each member of S contains all instances of *Dum*. From this we show, just as for the *Dum*-Lemma 8.14:

***Dum*-Lemma 9.12.** *Every non-last $R_{\circ}^{c}$-cluster is simple.*

Unwinding the Last Cluster

An R_o-*list* is a finite sequence $x_0, \ldots, x_n$ such that $x_i R_o x_{i+1}$ for all $i < n$. Now if C is the last R_o^c-cluster in our structure, then C may not be simple, i.e. may have more than one element. In that case, we will "unwind" C into a finite R_o-list. This can be done starting from any prescribed point $x \in C$, as follows. First pick some $y \in C$. Then $xR_o^c y$, so $xR_o^* y$, and so there is an R_o-list $x = x_0, \ldots, x_n = y$, with each x_i in C as C is last. If there is an element z of C that does not appear in this list, then since $x_n R_o^* z$ we can extend the list to $x_n R_o x_{n+1} R_o \cdots R_o x_k = z$, for some k. And so on. Eventually we build a finite R_o-list $x_0, \ldots, x_j$ in which every member of C appears at least once, and possibly more often. Since repetitions are allowed, we can arrange to end the list at any prescribed $z \in C$. Especially, we can arrange for the list to start and finish at the same point of C.

Now to define our state-sequence $\sigma : \omega \to S_\Gamma$. Let $C_0, \ldots, C_{n-1}$ be the sequence of non-last R_o^c-clusters in the order induced by R_o^c. Then for each $i < n$, from 9.12 it follows that C_i has the form $\{\sigma_i\}$ with $\sigma_i \in S_\Gamma$. This gives an R_o-list $\sigma_0 R_o \cdots R_o \sigma_{n-1}$. Then if C is the last R_o^c-cluster, there must be some $x \in C$ with $\sigma_{n-1} R_o x$. Let $\sigma_n = x$, and unwind C, as above, into an R_o-list $\sigma_n, \sigma_{n+1}, \ldots, \sigma_r$, that has $\sigma_n = \sigma_r$. Finally, we repeat this last list ad infinitum:

$$\sigma_n, \ldots, \sigma_r = \sigma_n, \sigma_{r+1} = \sigma_{n+1}, \ldots\ldots, \sigma_{r+q(r-n)+i} = \sigma_{n+i}, \ldots\ldots\ldots$$

(for all $q \in \omega$ and $0 \leq i < (r-n)$).

This completes the definition of σ. The main features of the construction are that for all $j \in \omega$,

(1) $\sigma_j R_o \sigma_{j+1}$: hence $\sigma_j R_o^* \sigma_k$ whenever $k \geq j$;

(2) if the R_o^c-cluster of x comes after that of σ_j, then $x = \sigma_k$ for some $k > j$; and

(3) if σ_j is in the last R_o^c-cluster, and so is x, then $x = \sigma_k$ for some $k > j$.

Theorem 9.13. *Let $\mathcal{M} = (S_\Gamma, \sigma, V_\Gamma)$. If $B \in \Gamma$, then for any $j \in \omega$ and $s \in \sigma_j$,*

$$B \in s \quad \text{iff} \quad \mathcal{M} \models_j B.$$

Proof.

For $B = p \in \Phi \cap \Gamma$, $\mathcal{M} \models_j p$ iff $\sigma_j \in V_\Gamma(p)$ iff $p \in s$. The truth-functional cases are straightforward as usual.

For the inductive case for $\bigcirc$, suppose $\bigcirc B \in s \cap \Gamma$, with $|s| = \sigma_j$. Pick any $t \in \sigma_{j+1}$. Then as $\sigma_j R_o \sigma_{j+1}$, the second filtration condition gives $B \in t$, whence the induction hypothesis on B gives $\mathcal{M} \models_{j+1} B$, so that $\mathcal{M} \models_j \bigcirc B$.

Conversely, if $\mathcal{M} \models_j \bigcirc B$, then $\mathcal{M} \models_{j+1} B$. Taking $\sigma_{j+1} = |t|$, we get $B \in t$ by the induction hypothesis, and so $\bigcirc B \in s$ by the *Fun*-Lemma 9.9.

Next the case of $\Box$. If $\Box B \in s \cap \Gamma$, then for any $k \geq j$, we have $\sigma_j R^*_\circ \sigma_k$, so that as $R^*_\circ$ is a Γ-filtration of $R^\Theta_\Box$, $B \in t$ for any $t \in \sigma_k$, hence $\mathcal{M} \models_k B$ by the induction hypothesis on B. This shows that $\mathcal{M} \models_j \Box B$.

On the other hand, if $\Box B \notin s$, then by the $R^c_\circ$-Lemma 9.11, $B \notin t$ for some t with $|s| R^c_\circ |t|$. If the $R^c_\circ$-cluster of $|s|$ is not last, then either $|t| = |s| = \sigma_j$, or else the $R^c_\circ$-cluster of $|t|$ comes after that of $|s|$, so by (2) above, $|t| = \sigma_k$ with $k > j$. If however $|s|$ is in the last $R^c_\circ$-cluster, then so too is $|t|$, so again, by (3), this gives $|t| = \sigma_k$ with $k > j$. In all cases then, we have $\mathcal{M} \not\models_k B$ for some $k \geq j$, and so $\mathcal{M} \not\models_j \Box B$.

Now we come to the case of the connective $\mathcal{U}$, and invoke for the first time the axioms $\mathcal{U}1$ and $\mathcal{U}2$. Make the inductive hypothesis that the Theorem holds for B and for D. Suppose that $B\mathcal{U}D \in s \cap \Gamma$, with $|s| = \sigma_j$. Then from $\mathcal{U}1$, $\Box\neg D \notin s$. Now the closure conditions on Γ give $\Box\neg D \in \Gamma$, and so from the hypothesis on D and the cases just treated, the Theorem holds for $\Box\neg D$. Hence $\mathcal{M} \not\models_j \Box\neg D$, implying that $\mathcal{M} \models_k D$ for some $k \geq j$. Take the least such k. If $k = j$, then immediately $\mathcal{M} \models_j B\mathcal{U}D$. Otherwise, when $k > j$, we will prove that $\mathcal{M} \models_i B$ for all i having $j \leq i < k$, again giving $\mathcal{M} \models_j B\mathcal{U}D$. The proof depends on showing that

$$t \in \sigma_i \quad \text{implies} \quad B\mathcal{U}D \in t. \qquad (\dagger)$$

This is done by induction on those i with $j \leq i < k$. For the case $i = j$, if $t \in \sigma_j$, then $B\mathcal{U}D \in t$ because $t \sim_\Gamma s$ and $B\mathcal{U}D \in s$. Now assume that (†) holds, with $i + 1 < k$. By the definition of k as "least", we then have $\mathcal{M} \not\models_i D$, so if $u \in \sigma_i$, it follows that $D \notin u$ by the hypothesis on D. But $B\mathcal{U}D \in u$, by (†), so applying axiom $\mathcal{U}2$ gives $\bigcirc(B\mathcal{U}D) \in u$. Since $\sigma_i R_\circ \sigma_{i+1}$ and $\bigcirc(B\mathcal{U}D) \in \Gamma$, any $t \in \sigma_{i+1}$ then has $B\mathcal{U}D \in t$ by the second filtration condition on $R_\circ$, establishing (†) for $i + 1$.

Thus (†) holds for all required i. Taking $t \in \sigma_i$ for such an i, the definition of k and hypothesis on D give $D \notin t$, so (†) and axiom $\mathcal{U}2$ then yield $B \in t$. Hence $\mathcal{M} \models_i B$, as desired, by hypothesis on B.

Conversely, suppose that $\mathcal{M} \models_j B\mathcal{U}D$ and $s \in \sigma_j$. Then for some $k \geq j$, $\mathcal{M} \models_k D$, with $\mathcal{M} \models_i B$ whenever $j \leq i < k$. We employ the thus-far unused implication of $\mathcal{U}2$ to show that the above condition (†) now holds whenever $j \leq i \leq k$. In particular, the case $i = j$ will then give our desideratum $B\mathcal{U}D \in s$.

The proof of (†) will this time go by backward induction on i. If $i = k$, then $D \in t$ for any $t \in \sigma_k$, by the hypothesis on D, and this immediately gives $B\mathcal{U}D \in t$ by $\mathcal{U}2$. Now assume that (†) holds with $j < i \leq k$. Then if $u \in \sigma_{i-1}$, (†) and the *Fun*-Lemma give $\bigcirc(B\mathcal{U}D) \in u$, since $\bigcirc(B\mathcal{U}D) \in \Gamma$

by definition of Γ, and $|u|R_{\text{o}}\sigma_i$. But $B \in u$, by hypothesis on B, so the other disjunct of $\mathcal{U}2$ applies to yeild $B\mathcal{U}D \in u$. Hence (†) holds for $i-1$.

This finishes the proof of Theorem 9.13.

To finish the completeness proof for Θ, recall that we began with a non-Θ-theorem A, and a point $s_A \in S$ with $A \notin s_A$. Taking a j such that $|s_A| = \sigma_j$, Theorem 9.13 gives $\mathcal{M} \not\models_j A$. Hence Θ is determined by the class of models on state sequences.

Finite Frame Property

Prima facie, our completeness proof for Θ does not yield the finite model property, since there are infinitely many sequences on any set with at least two elements, and we cannot effectively test a formula for truth at all points of a sequence. However we can rectify this by taking up the relational semantics in the form of the induction models introduced prior to the completeness proof. The idea is that instead of generating an infinite sequence, as above, we stop the R_{o}-list at

$$\sigma_0, \ldots, \sigma_n, \ldots, \sigma_r = \sigma_n$$

as soon as the last R_{o}^{c}-cluster is unwound. Then regarding all points in the list, except σ_r, and σ_n, as distinct, we get an induction frame in which A is invalid.

To formalise this, if r is any positive integer put

$$[0, r) = \{j \in \omega : 0 \leq j < r\},$$

and then for $0 \leq n < r$ define

$$\mathcal{F}_{n,r} = ([0, r), f),$$

where

$$f(j) = \begin{cases} j+1, & \text{if } j < r-1; \\ n, & \text{if } j = r-1. \end{cases}$$

Thus f is simply the successor function on $[0, r)$, except that the "successor" of the last element $r-1$ is n. We may visualise $\mathcal{F}_{n,r}$, as consisting of the initial segment $0, \ldots, n$, followed by the "simple loop" $n, n+1, \ldots, r-1, n$.

Now define a Φ_Γ-model $\mathcal{M}$ on $\mathcal{F}_{n,r}$, by putting

$$V(p) = \{j < r : \sigma_j \in V_\Gamma(p)\}.$$

Then by arguing as in the proof of Theorem 9.13, we can show that the statement of that theorem holds for this new model $\mathcal{M}$, provided $j < r$, where now $\mathcal{M} \models_j B$ means truth at the point j in the induction model, rather than truth at the state σ_j in a sequence model.

Hence Θ is determined by the class of finite induction frames $\mathcal{F}_{n,r}$.

Exercises 9.14

(1) Compute an upper bound on r for the induction frame $\mathcal{F}_{n,r}$ invalidating a prescribed non-Θ-theorem.

(2) Modify the state-sequence semantics to read

$$\begin{array}{lll} \mathcal{M} \models_j \Box A & \text{iff} & \text{for all } k > j,\ \mathcal{M} \models_k A \\ \mathcal{M} \models_j A\mathcal{U}B & \text{iff} & \text{for some } k > j,\ \mathcal{M} \models_k B \text{ and} \\ & & \mathcal{M} \models_i A \text{ whenever } j < i < k. \end{array}$$

Modify the given Θ-axioms to axiomatise the resulting set of valid formulae. (Do not introduce any essentially different axioms: deduce as a theorem the appropriate analogue of *Dum*.) Prove that this new logic is decidable.

Branching Time

The theory discussed so far has been concerned with logical properties of a *single* execution sequence $s_0, s_1, \ldots\ldots$ generated by processes acting in parallel. As mentioned at the outset, each state will have several possible successor states, and so there will be many different sequences that have a given starting state s_0. Thus any particular sequence will be but one "branch" of the "tree" of all possible future states. If we consider this tree as a whole, there a number of interesting new modal connectives that can be used to formalise reasoning about future behaviour:

$[\forall\mathrm{F}]A$:	along any future branch there is a state at which A is true, i.e. A is *inevitable*.
$[\exists\mathrm{F}]A$:	along some branch there is a state at which A is true, i.e. A is *potentially true*.
$[\forall\mathrm{G}]A$:	along all branches, A holds at all states, i.e. A is true at all possible future states.
$[\exists\mathrm{G}]A$:	along some branch, A holds at all states.
$[\forall\mathrm{X}]A$:	along every branch, A holds at the next state, i.e. A holds at all possible successor states.
$[\exists\mathrm{X}]A$:	A holds at some successor state.
$\forall(A\mathcal{U}B)$:	along every branch, it will be A until B.
$\exists(A\mathcal{U}B)$:	along some branch, it will be A until B.

A logical system embodying these notions, known as Computational Tree Logic (CTL), was introduced by Clarke and Emerson [1981,1982]. A similar system without the *until* operator was considered by Ben-Ari, Pnueli, and Manna [1983]. Emerson and Halpern [1985] established decidability and completeness for CTL, using a method of elimination of states from "psuedo-Hintikka structures". We will now see how to adapt their approach to the context of filtrations of canonical models.

Syntax and Semantics

The syntax for CTL is given by

$$A ::= p \mid \bot \mid A_1 \to A_2 \mid [\forall \mathrm{X}]A \mid \forall(A_1 \mathcal{U} A_2) \mid \exists(A_1 \mathcal{U} A_2)$$

The other connectives mentioned above are given by the following abbreviations.

$$\begin{array}{lll} [\forall \mathrm{F}]A & \text{is} & \forall(\top \mathcal{U} A) \\ [\exists \mathrm{F}]A & \text{is} & \exists(\top \mathcal{U} A) \\ [\forall \mathrm{G}]A & \text{is} & \neg\exists(\top \mathcal{U} \neg A) \\ [\exists \mathrm{G}]A & \text{is} & \neg\forall(\top \mathcal{U} \neg A) \\ [\exists \mathrm{X}]A & \text{is} & \neg[\forall \mathrm{X}]\neg A \end{array}$$

To define CTL-models, consider a frame $\mathcal{F} = (S, R)$ in which R is *serial*, i.e. $\forall s \exists t(sRt)$. Here sRt will be interpreted to mean that t is a *possible immediate successor* to s. An *R-branch starting at s* in $\mathcal{F}$ is an infinite sequence $s_0, \ldots, s_n, \ldots$ with $s = s_0$ and $s_n R s_{n+1}$ for all n. An *R-path* is a finite version of a branch, i.e. a sequence $s_0, \ldots, s_k$ with $s_n R s_{n+1}$ for all $n < k$. By seriality, any path extends to a branch.

Given the usual notion of a model $\mathcal{M} = (S, R, V)$ on such a frame, satisfaction of CTL-formulae is given by

$$\begin{array}{lll} \mathcal{M} \models_s [\forall \mathrm{X}]A & \text{iff} & \text{for all } t \in S,\ sRt \text{ implies } \mathcal{M} \models_t A. \\ \mathcal{M} \models_s \forall(A \mathcal{U} B) & \text{iff} & \text{for all } R\text{-branches } s = s_0 R s_1 R \cdots \\ & & \text{there exists } k \text{ with } \mathcal{M} \models_{s_k} B \text{ and} \\ & & \mathcal{M} \models_{s_i} A \text{ whenever } 0 \leq i < k. \\ \mathcal{M} \models_s \exists(A \mathcal{U} B) & \text{iff} & \text{for some } R\text{-branch } s = s_0 R s_1 R \cdots \\ & & \text{there exists } k \text{ with } \mathcal{M} \models_{s_k} B \text{ and} \\ & & \mathcal{M} \models_{s_i} A \text{ whenever } 0 \leq i < k. \end{array}$$

Axioms

Let CTL be the smallest logic in the language just described that contains the schemata

$$\begin{array}{ll} K_{\mathrm{X}}: & [\forall \mathrm{X}](A \to B) \to ([\forall \mathrm{X}]A \to [\forall \mathrm{X}]B) \\ D_{\mathrm{X}}: & [\exists \mathrm{X}]\top \\ \exists \mathcal{U}: & \exists(A \mathcal{U} B) \leftrightarrow (B \vee (A \wedge [\exists \mathrm{X}]\exists(A \mathcal{U} B))) \\ \forall \mathcal{U}: & \forall(A \mathcal{U} B) \leftrightarrow (B \vee (A \wedge [\forall \mathrm{X}]\forall(A \mathcal{U} B))) \end{array}$$

and is closed under Necessitation for $[\forall X]$ i.e.,

$$\vdash A \quad \text{implies} \quad \vdash [\forall X]A,$$

and under the following two rules:

$$\exists\text{-}Ind: \quad \vdash B \vee (A \wedge [\exists X]C) \rightarrow C$$
$$\text{implies}$$
$$\vdash \exists(A\mathcal{U}B) \rightarrow C,$$

$$\forall\text{-}Ind: \quad \vdash B \vee (A \wedge [\forall X]C) \rightarrow C$$
$$\text{implies}$$
$$\vdash \forall(A\mathcal{U}B) \rightarrow C.$$

Exercise 9.15

Show that CTL is sound with respect to the above semantics.

Completeness of CTL

We use the structure (S^C, R_X), where S^C is the set of CTL-maximal sets of formulae, and

$$sR_Xt \quad \text{iff} \quad \{B : [\forall X]B \in s\} \subseteq t.$$

If A_0 is a given non-theorem of CTL, there is some point $s_{A_0} \in S^C$ with $A_0 \notin s_{A_0}$. Let Γ be a *finite* set of formulae that has the following closure properties:

$A_0 \in \Gamma$;

Γ is closed under subformulae;

$\exists(A\mathcal{U}B) \in \Gamma$ implies $[\exists X]\exists(A\mathcal{U}B) \in \Gamma$;

$\forall(A\mathcal{U}B) \in \Gamma$ implies $[\forall X]\forall(A\mathcal{U}B) \in \Gamma$;

$[\exists X]\top \in \Gamma$,

and consider the structure (S_Γ, R), where S_Γ is defined as before (page 92), and R is the *least* Γ-filtration of R_X.

For $s \in S^C$, put $\Gamma_s = s \cap \Gamma$. For $x \in S_\Gamma$, put $\Gamma_x = \Gamma_s$ where s is some element of x.

Exercises 9.16

(1) Show that a finite Γ exists as described.

(2) Γ_x is well-defined (i.e. does not depend on the choice of $s \in x$).

(3) The relation R is *serial* on S_Γ.

Lemma 9.17. *Let $\exists(A\mathcal{U}B) \in \Gamma$. Then for any $x \in S_\Gamma$, $\exists(A\mathcal{U}B) \in \Gamma_x$ if, and only if, there exists in S_Γ an R-path $x = x_0, \ldots, x_k$ (i.e. x_iRx_{i+1} for all $i < k$) such that $A \in \Gamma_{x_i}$ for all $i < k$, and $B \in \Gamma_{x_k}$.*

Proof.

Suppose first that there is an R-path of the type described. Then we show that $\exists(A\mathcal{U}B) \in \Gamma_{x_i}$ for $0 \leq i < k$, by reverse induction on i, giving the desired conclusion when $i = 0$. We use the CTL-theorem (derived from axiom $\exists\mathcal{U}$)

$$(B \vee (A \wedge [\exists \mathrm{X}]\exists(A\mathcal{U}B))) \to \exists(A\mathcal{U}B).$$

For the base case $i = k$, we have $B \in \Gamma_{x_k}$ by assumption, so this CTL-theorem gives $\exists(A\mathcal{U}B) \in \Gamma_{x_k}$ by tautological consequence.

Now make the inductive hypothesis that $\exists(A\mathcal{U}B) \in \Gamma_{x_{i+1}}$. Choose $s \in x_i$ and $t \in x_{i+1}$ with $sR_{\mathrm{X}}t$ (since R is the *least* filtration of R_{X}). But $\exists(A\mathcal{U}B) \in t$, so $[\exists \mathrm{X}]\exists(A\mathcal{U}B) \in s$ by the second filtration condition (F2). Hence our CTL-theorem gives $\exists(A\mathcal{U}B) \in s$, and so $\exists(A\mathcal{U}B) \in \Gamma_{x_i}$ as desired.

For the converse direction, let X be the set of all points $x \in S_\Gamma$ for which there exists an R-path starting from x of the type described in the statement of the Lemma. We will show that whenever $\exists(A\mathcal{U}B) \in \Gamma_x$ then $x \in X$.

Now by the Definability Lemma 9.7, there is a formula C that is *characteristic for* X, i.e.

$$C \in s \quad \text{iff} \quad |s| \in X.$$

Let E be the formula

$$B \vee (A \wedge [\exists \mathrm{X}]C) \to C.$$

We show that E is a CTL-theorem, by showing that $E \in s$ for any $s \in S^C$.

First, if $B \vee (A \wedge [\exists \mathrm{X}]C) \notin s$, then it follows directly from the properties of s as a *maximal* set that $E \in s$ (Exercises 2.3). Thus we are reduced to the case that $B \vee (A \wedge [\exists \mathrm{X}]C) \in s$, and so either

$$B \in s, \tag{i}$$

or else

$$A \wedge [\exists \mathrm{X}]C \in s. \tag{ii}$$

Now if (i) holds, then putting $k = 0$ and $x = x_0 = |s|$ gives $B \in \Gamma_{x_k}$ and provides an R-path that makes $|s| \in X$. Thus $C \in s$, whence $E \in s$ by maximality of s (Ex. 2.3).

If, on the other hand, (ii) holds, then $[\exists \mathrm{X}]C \in s$, so there exists $t \in S^C$ such that $sR_{\mathrm{X}}t$ and $C \in t$. Then $|s|R|t|$ (by the first filtration condition (F1)) and $|t| \in X$, so there is an R-path of the desired type starting from $|t|$. But since $A \in \Gamma_{|s|}$ from (ii), appending $|s|$ to the beginning of this path gives a new R-path that ensures that $|s| \in X$, and so again $E \in s$.

This finishes the proof that E is a CTL-theorem. It then follows by the rule $\exists$-*Ind* that $\exists(A\mathcal{U}B) \to C$ is a CTL-theorem, and so belongs to every CTL-maximal set. Thus for any $s \in S^C$, if $\exists(A\mathcal{U}B) \in \Gamma_{|s|}$ then $\exists(A\mathcal{U}B) \in s$, so $C \in s$, giving $|s| \in X$ as desired to complete the proof of Lemma 9.17.

Exercise 9.18

Let $\forall(A\mathcal{U}B) \in \Gamma$. Show that if $\forall(A\mathcal{U}B) \notin \Gamma_x$, then there exists an R-branch $x = x_0, \dots, x_k, \dots$ such that for no k do we have $B \in \Gamma_{x_k}$ simultaneously with $A \in \Gamma_{x_i}$ for all $i < k$.

If the converse of Exercise 9.18 were true, then in combination with Lemma 9.17 we would obtain a Filtration Lemma for the model (S_Γ, R, V_Γ) similar to Theorem 9.13, and completeness for CTL would follow. However it could be that while $\forall(A\mathcal{U}B) \in \Gamma_x$, a branch of the type described in 9.18 nonetheless exists to prevent $\forall(A\mathcal{U}B)$ being "true" at x. We are going to have to "unravel" R to get around this, and the structure we use for this unravelling is a special type of *tree*.

Γ-Trees

Let (T, ρ) be a frame with ρ irreflexive. The members of T will be called *nodes*. If $n\rho m$, then m is a *successor* of n, while n is a *predecessor* of m. The frame will be called a *tree* if each of its nodes has at most one predecessor.

A tree is *rooted* if it has a unique element r, the *root*, that generates it, i.e. has $T = \{m : r\rho^* m\}$. Note that for each node m, there will be a unique ρ-path from the root r to m.

A *leaf* in a tree is a node that has no successors. Non-leaf nodes are said to be *interior*.

For finite trees, the word "*branch*" will be used in a modified way to mean a path (i.e. a ρ-path) whose last node is a leaf.

We will work with trees who nodes are *labelled* by members of S_Γ, i.e. there is a function assigning to each $n \in T$ a label $\underline{n} \in S_\Gamma$. Then a formula B is said to be *realised at n* if $B \in \Gamma_{\underline{n}}$, while B *fails at n* if $B \notin \Gamma_{\underline{n}}$.

A *Γ-tree* is a finite rooted tree who nodes are labelled by member of S_Γ in such a way that

(Γ1) if m is a successor node to n, then $\underline{n}R\underline{m}$;

(Γ2) if $[\forall X]A \in \Gamma$, and $[\forall X]A$ fails at an interior node n, then A fails at some successor node of n.

A Γ-tree is *rooted at* $x \in S_\Gamma$ if x is the label of its root node.

Exercise 9.19

If $[\exists X]A$ belongs to Γ and is realised at an interior node of a Γ-tree, then A is realised at a successor of that node.

Lemma 9.20. *Let $\forall(A\mathcal{U}B) \in \Gamma$. Then if $\forall(A\mathcal{U}B) \in \Gamma_x$, there is a Γ-tree T rooted at x such that A is realised at every interior node of T, and B is realised at every leaf.*

Proof.

Let X be the set of points $x \in S_\Gamma$ for which there exists a tree rooted at x with the properties described in the statement of the Lemma. Let C be a formula that is characteristic for X, i.e.

$$C \in s \quad \text{iff} \quad |s| \in X.$$

Then it is enough to show that

$$\forall(A\mathcal{U}B) \to C$$

is a CTL theorem. Hence by rule $\forall$-*Ind*, it is enough to show that the formula

$$B \vee (A \wedge [\forall X]C) \to C \tag{E}$$

belongs to every $s \in S^C$.

The reasoning is like that for the proof of Lemma 9.17. First, if the formula $B \vee (A \wedge [\forall X]C)$ is not in s, then $E \in s$ follows directly. Thus we are reduced to the case that either

$$B \in s, \tag{i}$$

or else

$$A \wedge [\forall X]C \in s. \tag{ii}$$

Now if (i) holds, then putting $T = \{r\}$ with $r = |s|$ and $\underline{r} = r$ gives a one-node Γ-tree demonstrating that $|s| \in X$, so $C \in s$, and hence $E \in s$.

Suppose, on the other hand, that (ii) holds. Let $[\forall X]D_1, \ldots, [\forall X]D_k$ be all formulae in Γ of the form $[\forall X]D$ that do not belong to s. For each i with $1 \leq i \leq k$ there is some $t_i \in S^C$ such that $sR_X t_i$ and $D_i \notin t_i$. Then $|s|R|t_i|$ and $D_i \notin \Gamma_{|t_i|}$. Construct a Γ-tree consisting of a root r labelled by $|s|$, with k successors $m_1, \ldots, m_k$ having m_i labelled by $|t_i|$. The only interior node is r, and this has $A \in \Gamma_{\underline{r}}$, i.e. A is realised at r, since $A \in s$ by (ii).

Now extend this tree as follows. For each i, since (ii) gives $[\forall X]C \in s$, we have $C \in t_i$, and so $|t_i| \in X$. Hence there is a Γ-tree T_i rooted at $|t_i|$ that has A realised at all its interior nodes and B realised at all its leaves. Replace m_i by the tree T_i, i.e. identify m_i with the root of T_i.

The effect of this construction is to create a Γ-tree rooted at $|s|$ that makes $|s| \in X$, leading to $E \in s$ as desired, and completing the proof.

Fulfilment

Let T be a Γ-tree.

- A path in T *realises* $A\mathcal{U}B$ if there is a node on the path at which B is realised, while all earlier nodes on the path realise A.
- The formula $\exists(A\mathcal{U}B)$ is *fulfilled* at node n in T if either it fails at n, i.e. $\exists(A\mathcal{U}B) \notin \Gamma_{\underline{n}}$, or else there exists a path in T that starts from n and realises $A\mathcal{U}B$.
- The formula $\forall(A\mathcal{U}B)$ is *fulfilled* at node n in T if either it fails at n, or else every branch in T that starts from n realises $A\mathcal{U}B$.

Formulae of the form $\exists(A\mathcal{U}B)$ and $\forall(A\mathcal{U}B)$ will be called *eventuality formulae.*

Fulfilment Lemma 9.21. *Let T be a Γ-tree, and C an eventuality formula in Γ that is not fulfilled at node n in T.*
(1) *If $C = \exists(A\mathcal{U}B)$, then there is a branch in T starting from n with A and $\exists(A\mathcal{U}B)$ realised at all nodes of the branch.*
(2) *If $C = \forall(A\mathcal{U}B)$, then for every branch in T starting from n, either the branch realises $A\mathcal{U}B$, or else A and $\forall(A\mathcal{U}B)$ are realised at all nodes of the branch.*

Proof.
(1). Since $\exists(A\mathcal{U}B)$ is not fulfilled at n, $\exists(A\mathcal{U}B) \in \Gamma_{\underline{n}}$. Moreover, if B were realised at n, then $\exists(A\mathcal{U}B)$ would be fulfilled there, contrary to hypothesis. Hence $B \notin \Gamma_{\underline{n}}$. But the formula

$$\exists(A\mathcal{U}B) \to (B \lor (A \land [\exists X]\exists(A\mathcal{U}B)))$$

is CTL-derivable from axiom $\exists\mathcal{U}$, and $[\exists X]\exists(A\mathcal{U}B) \in \Gamma$, so it follows that both A and $[\exists X]\exists(A\mathcal{U}B)$ are realised at n.

Now if n is an interior node, then by Exercise 9.19 it has a successor node m that realises $\exists(A\mathcal{U}B)$. Since $\exists(A\mathcal{U}B)$ is not fulfilled at n while A is realised at n, B cannot be realised at m, so the above argument applies to give that both A and $[\exists X]\exists(A\mathcal{U}B)$ are realised at m. If m in turn is interior, the construction repeats, generating a path of the desired kind, and stopping only when a leaf is reached.

Exercise 9.22

Prove part (2) of Fulfilment Lemma 9.21.

Theorem 9.23. *For any $x \in S_\Gamma$ there exists a Γ-tree T_x with root r labelled by x, such that*
(1) *if $[\forall X]A \in \Gamma$ and $[\forall X]A$ fails at r, then A fails at some successor of r;*

(2) *every eventuality formula in Γ is fulfilled at r.*

Proof.

First Stage: Construct a Γ-tree by taking a root node r, labelled by x, and for each formula $[\forall X]A$ in Γ that fails at r, take some $y \in S_\Gamma$ with xRy and $A \notin \Gamma_y$, and add a successor node to r labelled by y. This ensures already that (1) holds.

Next make a series of extensions to the tree to establish (2), at each stage adding new nodes or sub-trees below the leaves of the tree thus far created (the reader should be visualising trees as growing downwards). It follows that at all stages r continues to be the root of the tree being constructed.

To see how this works, let T be the tree that has been created at some stage, and suppose C is an eventuality formula in Γ that is not fulfilled at r in T.

Case of $\exists\mathcal{U}$: If $C = \exists(A\mathcal{U}B)$, then by (1) of Fulfilment Lemma 9.21 there is a branch in T from r to a leaf m having $\exists(A\mathcal{U}B)$ and A realised at every node. By Lemma 9.17 there is an R-path $\underline{m} = x_0, \ldots, x_k$ in S_Γ with B realised at x_k, and A realised at x_i for $0 \leq i < k$. Extend T by adjoining a path $m_0, \ldots, m_k$ of nodes with $m = m_0$, and put $\underline{m_i} = x_i$ for $0 \leq i \leq k$. Then for each $i < k$, repeat the First Stage construction to adjoin enough successor nodes to m_i to ensure that whenever a $[\forall X]D$-type formula from Γ fails at m_i, then D fails at a successor of m_i. When this is done, we have a new Γ-tree with $\exists(A\mathcal{U}B)$ fulfilled at the root r.

Case of $\forall\mathcal{U}$: If $C = \forall(A\mathcal{U}B)$, proceed as follows. Let m be any leaf of T and consider the branch from r to m in T. If this branch realises $A\mathcal{U}B$, leave m alone. Otherwise, by (2) of 9.21, A and $\forall(A\mathcal{U}B)$ are realised at every node of the branch. Hence $\forall(A\mathcal{U}B) \in \Gamma_{\underline{m}}$, so by Lemma 9.20 there is is a Γ-tree $T_{\underline{m}}$ rooted at $\underline{m}$ with A realised at every interior node of $T_{\underline{m}}$, and B realised at every leaf. Adjoin this tree, by identifying m with the root of $T_{\underline{m}}$. The result is a structure in which every branch passing through m realises $A\mathcal{U}B$.

By applying this procedure to each leaf m of T, we end up with a Γ-tree fulfilling $\forall(A\mathcal{U}B)$ at r.

Notice that once a formula $\exists(A\mathcal{U}B)$ becomes fulfilled at r, it remains so if any new nodes are added. But the same is true for a formula $\forall(A\mathcal{U}B)$, because of the crucial fact that in each extension the new nodes are always added *below* an old leaf. Thus any branch from r in the new tree must be an extension of a branch from r in the old tree, so that if all the old branches realise $\forall(A\mathcal{U}B)$, then all the new ones will as well.

The upshot of all this is that by making finitely many repetitions of these constructions, a tree will be produced in which all eventuality formulae from Γ are fulfilled at r.

Final Model

We are now at the final stage of our construction of a finite CTL-model. This is done by joining together copies of the trees described in Theorem 9.23 (the result need not itself be a tree: it may contain cycles).

Begin with the tree T_x of any $x \in S_\Gamma$, as provided by 9.23. Replace each leaf m of T_x by the tree $T_{\underline{m}}$ (i.e. identify m with the root of $T_{\underline{m}}$). Repeat this process for the leaves of the newly adjoined trees, except in the case of a leaf n for which $T_{\underline{n}}$ has already been adjoined. In this case, delete n and draw an edge from the predecessor of n to the root of $T_{\underline{n}}$, i.e. make the root of $T_{\underline{n}}$ a successor of the predecessor of n (this is the part of the construction that may introduce cycles).

Since each tree T_x is finite, and there are finitely many labels $x \in S_\Gamma$, this process terminates in the construction of a finite frame (T, ρ) whose points are all labelled by members of S_Γ, and whose relation is given by the successor relation on the trees T_x. Now make this frame into a model $\mathcal{N}$ by putting

$$\mathcal{N} \models_n p \quad \text{iff} \quad p \in \Gamma_{\underline{n}}.$$

Exercise 9.24

Use the fact that $[\exists \mathrm{X}]\top \in \Gamma$ to prove that ρ is serial.

Theorem 9.25. *If $A \in \Gamma$, then for any node $n \in T$,*

$$\mathcal{N} \models_n A \quad \text{iff} \quad A \in \Gamma_{\underline{n}}.$$

Proof. We consider only the major inductive cases.

$[\forall \mathrm{X}]$-*Case*: Suppose the result holds for A, and $[\forall \mathrm{X}]A \in \Gamma$. If $\mathcal{N} \not\models_n [\forall \mathrm{X}]A$, then $\mathcal{N} \not\models_m A$ for some successor m of n. Then by the definition of Γ-trees, $\underline{n}R\underline{m}$, while $A \notin \Gamma_{\underline{m}}$ by the induction hypothesis, so as R is a Γ-filtration of R_{X}, $[\forall \mathrm{X}]A \notin \Gamma_{\underline{n}}$.

Conversely, if $[\forall \mathrm{X}]A \notin \Gamma_{\underline{n}}$, the definition of Γ-trees (when n is interior) and 9.23(1) (when n is a root) ensure that A fails at some successor of n, making $[\forall \mathrm{X}]A$ false in $\mathcal{N}$ at n by the induction hypothesis.

$\exists\mathcal{U}$-*Case*: Suppose the result holds for A and B, and $\exists(A\mathcal{U}B) \in \Gamma$.

If $\mathcal{N} \models_n \exists(A\mathcal{U}B)$, then there is a ρ-path $n = n_0, \ldots, n_k$ such that, by the induction hypothesis, B is realised at n_k, and A realised at n_i for all $i < k$. But then $\underline{n_0}, \ldots, \underline{n_k}$ is an R-path in S_Γ, so Lemma 9.17 gives $\exists(A\mathcal{U}B) \in \Gamma_{\underline{n}}$.

Conversely, let $\exists(A\mathcal{U}B) \in \Gamma_{\underline{n}}$. Suppose $n \in T_x$. Now either $A\mathcal{U}B$ is realised in T_x by a path starting at n, or else by 9.21(1) there is a branch from n to a leaf m of T_x with A and $\exists(A\mathcal{U}B)$ realised at all nodes, including m. But in that case, since $\exists(A\mathcal{U}B)$ is fulfilled at the root of $T_{\underline{m}}$ (9.23(2)) there must be a path from the root in $T_{\underline{m}}$ that realises $A\mathcal{U}B$.

In either case, we get a path in T, starting from n, that realises $A\mathcal{U}B$. Hence by the induction hypothesis, this path makes $\mathcal{N} \models_n \exists(A\mathcal{U}B)$.

$\forall\mathcal{U}$-Case: Suppose that $\forall(A\mathcal{U}B) \in \Gamma_{\underline{n}}$, where n is a node in some T_x. Consider any ρ-branch from $\underline{n}$ in T. Then there must be a leaf m of T_x such that this branch passes through the root of $T_{\underline{m}}$. Now either $A\mathcal{U}B$ is realised by the path from n to m in T_x, or else by 9.21(2) A and $\forall(A\mathcal{U}B)$ are realised by all nodes of this path. But the ρ-branch must pass through $T_{\underline{m}}$, and so by (9.23(2)), $A\mathcal{U}B$ will be realised along the part of the ρ-branch that lies in $T_{\underline{m}}$, and hence be realised along the ρ-branch itself. With the induction hypothesis, this shows that $\mathcal{N} \models_n \forall(A\mathcal{U}B)$.

For the converse, suppose that $\forall(A\mathcal{U}B) \notin \Gamma_{\underline{n}}$. To prove that $\mathcal{N} \not\models_n \forall(A\mathcal{U}B)$, we use the one part of the axiomatisation of CTL that has yet to play a role: the implication

$$(B \vee (A \wedge [\forall X]\forall(A\mathcal{U}B))) \to \forall(A\mathcal{U}B) \tag{\dagger}$$

that is part of axiom $\forall\mathcal{U}$. Since $\forall(A\mathcal{U}B) \notin \Gamma_{\underline{n}}$, this immediately yields $B \notin \Gamma_{\underline{n}}$, and hence $\mathcal{N} \not\models_n B$ by induction hypothesis.

Now if $A \notin \Gamma_{\underline{n}}$, then $\mathcal{N} \not\models_n A$, so as $\mathcal{N} \not\models_n B$ we have $\mathcal{N} \not\models_n \forall(A\mathcal{U}B)$ immediately. If, on the other hand, $A \in \Gamma_{\underline{n}}$, (†) yields $[\forall X]\forall(A\mathcal{U}B) \notin \Gamma_{\underline{n}}$. But $[\forall X]\forall(A\mathcal{U}B) \in \Gamma$, so by the definition of Γ-trees (when n is interior) and 9.23(1) (when n is a root), there must be a ρ-successor n_1 of n with $\forall(A\mathcal{U}B) \notin \Gamma_{\underline{n_1}}$.

The argument now repeats itself: if $A \notin \Gamma_{\underline{n_1}}$, then $\mathcal{N} \not\models_{n_1} A$, so $\mathcal{N} \not\models_{n_1} \forall(A\mathcal{U}B)$ as $B \notin \Gamma_{n_1}$ and hence $\mathcal{N} \not\models_{n_1} B$. If $A \in \Gamma_{\underline{n_1}}$, then there is a successor n_2 of n_1 with $\forall(A\mathcal{U}B) \notin \Gamma_{\underline{n_2}}$, and so on. The argument either generates a ρ-path $n = n_0, \ldots, n_k$ with $\mathcal{N} \not\models_{n_k} A$ and $\mathcal{N} \not\models_{n_i} B$ for all $i \leq k$, or else it generates a ρ-branch $n = n_0, \ldots, n_k, \ldots$ with $\mathcal{N} \not\models_{n_i} B$ for all i. In either case, it follows that $\mathcal{N} \not\models_n \forall(A\mathcal{U}B)$.

This completes our discussion of the proof of Theorem 9.25.

Exercise 9.26

Finish the argument showing that CTL has the finite model property and is decidable.

10 | Propositional Dynamic Logic

Dynamic logic (Pratt [1976]) is based on the idea of associating with each command α of a programming language a modal connective $[\alpha]$, with the formula $[\alpha]A$ being read "after α terminates, A", i.e. "after every terminating execution of α, A is true" (allowing that a non-deterministic α may be executed in more than one way). The dual formula $<\alpha>A$ then means "there is an execution of α that terminates with A true" (recall the discussion of motivations in §1).

In this way we obtain a multi-modal language, with a set of modal connectives indexed by the set of programs. An interesting theory emerges about the ways in which properties of complex programs can be expressed by the modal connectives of their constituent programs. The programs themselves are generated from some set Π of "atomic" programs, whose nature is not examined further, so that we can concentrate on the behaviour of operations that generate new commands from given ones. Thus Π plays the same role for programs that Φ plays for formulae of propositional logic. What happens when we replace Π by actual commands will be the subject of Part Three.

Syntax

Atomic formulae:	$p \in \Phi$
Atomic programs:	$\pi \in \Pi$
Formulae:	$A \in Fma(\Phi, \Pi)$
Programs:	$\alpha \in Prog(\Phi, \Pi)$

$$A ::= p \mid \bot \mid A_1 \to A_2 \mid [\alpha]A$$

$$\alpha ::= \pi \mid \alpha_1; \alpha_2 \mid \alpha_1 \cup \alpha_2 \mid \alpha^* \mid A?$$

Intended meanings are:

$[\alpha]A$	after α, A,
$\alpha_1; \alpha_2$	do α_1 and then α_2 (*composition*),
$\alpha_1 \cup \alpha_2$	do either α_1 or α_2 non-deterministically (*alternation*),
α^*	repeat α some finite number (≥ 0) of times (*iteration*),
$A?$	*test* A: continue if A is true, otherwise "fail".

Further constructs are introduced by definitional abbreviation:

$<\alpha>A$	is	$\neg[\alpha]\neg A$,
if A **then** α **else** β	is	$(A?;\alpha)\cup(\neg A?;\beta)$
while A **do** α	is	$(A?;\alpha)^*;\neg A?$
repeat α **until** A	is	$\alpha;(\neg A?;\alpha)^*$
skip	is	$\top?$
abort	is	$\bot?$
α^0	is	**skip**
α^{n+1}	is	$(\alpha;\alpha^n)$

Standard Models

According to §5, a model for the language just described should be a structure of the form

$$\mathcal{M} = (S, \{R_\alpha : \alpha \in Prog(\Phi, \Pi)\}, V),$$

with R_α a binary relation on S for each program α, and

$$\mathcal{M} \models_s [\alpha]A \quad \text{iff} \quad sR_\alpha t \text{ implies } \mathcal{M} \models_t A.$$

We want the binary relations R_α to reflect the intended meanings of programs α. Thus a model $\mathcal{M}$ will be defined to be *standard* if it satisfies the following conditions:

$$\begin{aligned} R_{\alpha;\beta} &= R_\alpha \circ R_\beta = \{(s,t) : \exists u(sR_\alpha u \ \&\ uR_\beta t)\}; \\ R_{\alpha\cup\beta} &= R_\alpha \cup R_\beta; \\ R_{\alpha^*} &= R_\alpha^* = \text{ ancestral of } R_\alpha; \\ R_{A?} &= \{(s,s) : \mathcal{M} \models_s A\}. \end{aligned}$$

There are no constraints on the R_π's. This means that given a structure

$$(S, \{R_\pi : \pi \in \Pi\}, V)$$

which assigns a binary relation to each atomic program, a uniquely determined standard model is obtained by using the above standard model conditions to inductively *define* R_α for non-atomic programs α.

Exercises 10.1

(1) In a standard model $\mathcal{M}$, show:
   (i) $R_{\mathbf{skip}} = \{(s,s) : s \in S\}$;
   (ii) $R_{\mathbf{abort}} = \emptyset$;

(iii) $A?$ has the same meaning as

if A **then skip else abort**;

(iv) $R_{\alpha^n} = (R_\alpha)^n$;

(v) $\mathcal{M} \models [\alpha^n]A \leftrightarrow [\alpha]^n A$ (recall the definition of $\Box^n$ from Exercise 3.9(6));

(vi) $\mathcal{M} \models_s [\alpha^*]A$ iff for all $n \geq 0$, $\mathcal{M} \models_s [\alpha^n]A$.

(2) In a standard model, determine the nature of R_α when α is a while-command (**while** A **do** α_1), or a conditional command (**if** A **then** α_1 **else** α_2).

(3) Formulate precisely the observation that in a standard model, any execution of a program consists of a finite sequence of "atomic executions".

Axioms

Let PDL be the smallest normal logic in $Fma(\Phi, \Pi)$ that contains the schemata

Comp:	$[\alpha;\beta]A \leftrightarrow [\alpha][\beta]A$,
Alt:	$[\alpha \cup \beta]A \leftrightarrow [\alpha]A \wedge [\beta]A$,
Mix:	$[\alpha^*]A \rightarrow A \wedge [\alpha][\alpha^*]A$,
Ind:	$[\alpha^*](A \rightarrow [\alpha]A) \rightarrow (A \rightarrow [\alpha^*]A)$,
Test:	$[A?]B \leftrightarrow (A \rightarrow B)$.

Notice the correspondence between $[\alpha^*]$ and $[\alpha]$ in the present language, and $\Box$ and $\bigcirc$ in temporal logic. The axioms *Mix* and *Ind* here correspond exactly to the axioms with the same names in §9. This is to be expected, since in each case, one connective is interpreted as the ancestral of the interpretation of the other.

We will show that PDL is determined by, and has the finite model property with respect to, the class of *standard* models.

Exercises 10.2

(1) $\vdash_{PDL} [\alpha^n]A \leftrightarrow [\alpha]^n A$.

(2) $\vdash_{PDL} [\alpha^*]A \rightarrow [\alpha]^n A$.

(3) (*Soundness*) If $\vdash_{PDL} A$, then A is true in all standard models.

Completeness of PDL

Let $\mathcal{M}^P = (S^P, \{R^P_\alpha : \alpha \in Prog(\Phi,\Pi)\}, V^P)$ be the canonical PDL-model, with S^P the set of PDL-maximal sets,

$$sR^P_\alpha t \quad \text{iff} \quad \{B : [\alpha]B \in s\} \subseteq t,$$

and

$$V^P(p) = \{s \in S^P : p \in S\}.$$

Although $\mathcal{M}^P$ verifies all PDL-theorems, and falsifies all non-theorems, it has the same inadequacy that occurred with the temporal logic of §9: $R^P_{\alpha^*}$ is not the ancestral of R^P_α. However we do have:

Theorem 10.3. *$\mathcal{M}^P$ satisfies all standard-model conditions except*

$$R^P_{\alpha^*} \subseteq (R^P_\alpha)^*.$$

Proof. We discuss briefly only part of one condition, namely,

$$R^P_{\alpha;\beta} \subseteq R^P_\alpha \circ R^P_\beta.$$

Suppose $sR^P_{\alpha;\beta}t$. We need to find a $u \in S^P$ with $sR^P_\alpha u$ and $uR^P_\beta t$. It suffices, by Lindenbaum's Lemma, to show that

$$u_0 = \{B : [\alpha]B \in s\} \cup \{\neg[\beta]D : D \notin t\}$$

is PDL-consistent, and for this the PDL-theorem

$$[\alpha][\beta]A \to [\alpha;\beta]A$$

is used. The proof is very similar to the use of the schema

$$\Box\Box A \to \Box A$$

in Theorem 3.6 to derive the weak density condition.

Exercise 10.4

Complete the proof of Theorem 10.3.

(The completeness theorem to follow will not depend on 10.3.)

Now let A be a fixed non-theorem of PDL. To obtain a *standard* model that falsifies A we will collapse $\mathcal{M}^P$ by a suitable Γ that contains A. The closure rules for Γ that will be needed are:

Γ is closed under subformulae;
$[B?]D \in \Gamma$ implies $B \in \Gamma$;
$[\alpha;\beta]B \in \Gamma$ implies $[\alpha][\beta]B \in \Gamma$;
$[\alpha \cup \beta]B \in \Gamma$ implies $[\alpha]B, [\beta]B \in \Gamma$;
$[\alpha^*]B \in \Gamma$ implies $[\alpha][\alpha^*]B \in \Gamma$.

A set Γ satisfying these conditions will be called *closed*.

Lemma 10.5. (Fischer and Ladner [1979]). *If Γ is the smallest closed set containing a given formula A, then Γ is finite.*

Proof. The point is to show that closing $Sf(A)$ under the above rules produces only finitely many new formulae. Define a formula to be *boxed* if it is prefixed by a modal connective, i.e. is of the form $[\alpha]B$ for some α and B. Each time we apply a closure rule, new boxed formulae appear on the right side of the rule, and further rules may apply to these new formulae. But observe that the programs α indexing prefixes $[\alpha]$ on the right side are in all cases shorter in length than those indexing the prefix on the left of the rule in question. Hence we will eventually produce only atomic prefixes on the right, and run out of rules to apply.

To formalise this argument, define a formula Y to be a *derivative* of formula X, denoted $X \succ Y$, if one of the following obtains:

X is $[B?]D$, and Y is a subformula of B;
X is $[\alpha;\beta]B$, and Y is $[\alpha][\beta]B$ or $[\beta]B$;
X is $[\alpha \cup \beta]B$, and Y is $[\alpha]B$ or $[\beta]B$;
X is $[\alpha^*]B$, and Y is $[\alpha][\alpha^*]B$.

Then the smallest closed set Γ containing A is obtained by closing $Sf(A)$ under $\succ$, i.e. $D \in \Gamma$ iff there is a finite sequence of the form

$$X = X_0 \succ \cdots\cdots \succ X_n = D,$$

with $X \in Sf(A)$. Notice that the definition of $\succ$ is arranged so that if a set Δ of formulae is closed under subformulae, then adding to Δ all the derivatives of some $X \in \Delta$ will result in a set still closed under subformulae. Thus to see that Γ is finite, observe that only boxed formulae have derivatives, and if $[\alpha]B \succ [\beta]D$, then the length of β as a string of symbols is less than that of α, so that there can be no infinitely-long $\succ$-sequences. Since $Sf(A)$ is finite, and each formula has only finitely many derivatives, it follows that only finitely many formulae result by forming $\succ$-sequences originating in $Sf(A)$.

Having determined that Γ, the smallest closed set containing A, is finite, we perform a Γ-filtration of $\mathcal{M}^P$. Let Φ_Γ be $\Phi \cap \Gamma$, and let $Prog_\Gamma$ be the smallest set of programs that includes

all atomic programs occurring in members of Γ, and
all tests $B?$ occurring in members of Γ,

and is closed under ;, $\cup$, and *. Define a model

$$\mathcal{M}_\Gamma = (S_\Gamma, \{R_\alpha : \alpha \in Prog_\Gamma\}, V_\Gamma),$$

where S_Γ and V_Γ are as usual, while R_π is *any* Γ-filtration of R^P_π,

$$R_{B?} = \{(|s|, |s|) : \mathcal{M}^P \models_s B\},$$

and otherwise R_α is given inductively by the standard-model condition on α.

Exercise 10.6

Show that if $B?$ occurs in Γ, then $B \in \Gamma$, and hence that $R_{B?}$ is well defined.

Theorem 10.7. *$\mathcal{M}_\Gamma$ is a Γ-filtration of $\mathcal{M}^P$.*

Proof.
We have to show that R_α is a Γ-filtration of R^P_α whenever $\alpha \in Prog_\Gamma$. The case of atomic α holds by definition.

Tests. Suppose $B? \in Prog_\Gamma$. Let $sR^P_{B?}t$. Then if $D \in s$, $(B \to D) \in s$, so $[B?]D \in s$ by axiom $Test$, hence $D \in t$. Thus $s \subseteq t$, and therefore $s = t$ as s is maximal (2.3(2)). Moreover, as $Test$ implies $\vdash_{PDL} [B?]B$, we get $B \in t = s$. Thus we have $s = t$ and $\mathcal{M}^P \models_s B$, implying $|s|R_{B?}|t|$ by definition of $R_{B?}$. Hence (F1) holds for $B?$.

For the second filtration condition, suppose that $|s|R_{B?}|t|$. Then $|s| = |t|$ and $\mathcal{M}^P \models_s B$. Thus if $[B?]D \in \Gamma$ and $\mathcal{M}^P \models_s [B?]D$, we have $\mathcal{M}^P \models_s (B \to D)$, as $\mathcal{M}^P \models Test$, and so $\mathcal{M}^P \models_s D$. But then $\mathcal{M}^P \models_t D$, since $s \sim_\Gamma t$ and $D \in \Gamma$.

This completes the proof that $R_{B?}$ is a Γ-filtration of $R^P_{B?}$.

The proof of the first filtration condition (F1) in the inductive cases will use the following idea (which was used in the Ancestral Lemma 9.8). Given $s \in S^P$, let A_s be a formula having

$$A_s \in t \quad \text{iff} \quad |s|R_\alpha|t|$$

(A_s exists by Definability Lemma 9.7). Then to show that

$$sR^P_\alpha t \quad \text{implies} \quad |s|R_\alpha|t|,$$

it suffices to prove that $[\alpha]A_s \in s$, for then if $sR^P_\alpha t$ we get $A_s \in t$ as desired.

Composition. Suppose that $(\alpha;\beta) \in Prog_\Gamma$, and, inductively, that R_α and R_β are Γ-filtrations of R^P_α and R^P_β, respectively.

Let A_s be a formula having

$$A_s \in t \quad \text{iff} \quad |s|R_{\alpha;\beta}|t|.$$

Now if $sR^P_\alpha uR^P_\beta t$, then by the induction hypothesis $|s|R_\alpha|u|R_\beta|t|$, hence $|s|R_{\alpha;\beta}|t|$ as $\mathcal{M}_\Gamma$ is standard for $(\alpha;\beta)$, and so $A_s \in t$. This shows that $[\alpha][\beta]A_s \in s$, and so by axiom *Comp*, $[\alpha;\beta]A_s \in s$ as needed to ensure that $sR^P_{\alpha;\beta}t$ implies $|s|R_{\alpha;\beta}|t|$.

If $|s|R_{\alpha;\beta}|t|$ then for some u, $|s|R_\alpha|u|$ and $|u|R_\beta|t|$. Then if formula $[\alpha;\beta]B$ is in Γ and true at s in $\mathcal{M}^P$, $[\alpha][\beta]B$ is true at s, as $\mathcal{M} \models Comp$, and also a member of Γ by a closure condition. But then the hypotheses on α and β give $[\beta]B$ true at u in $\mathcal{M}^P$, and thence B true at t.

Alternation. The inductive case for $(\alpha \cup \beta)$ is similar to that for $(\alpha;\beta)$. If A_s is a formula having

$$A_s \in t \quad \text{iff} \quad |s|R_{\alpha\cup\beta}|t|,$$

then using the inductive hypothesis on α and β, and the fact that $\mathcal{M}_\Gamma$ is standard for $(\alpha \cup \beta)$, we get $A_s \in t$ whenever $sR^P_\alpha t$ or $sR^P_\beta t$. Hence $[\alpha]A_s, [\beta]A_s \in s$, so $[\alpha \cup \beta]A_s \in s$ by axiom *Alt*.

The proof that $R_{\alpha\cup\beta}$ satisfies (F2) is left as an exercise.

Iteration. The proof that R_{α^*}, i.e. R^*_α, satisfies (F1) in relation to $R^P_{\alpha^*}$ is exactly the same as the proof of the Ancestral Lemma 9.8., using *Ind*. For (F2), we need to show that

if $|s|R_{\alpha^*}|t|$, then for all B,
  if $[\alpha^*]B \in \Gamma$ and $\mathcal{M}^P \models_s [\alpha^*]B$, then $\mathcal{M}^P \models_t B$.

But if R_α is a Γ-filtration of R^P_α, we can show that for all $n \geq 0$,

if $|s|R^n_\alpha|t|$, then for all B,
  if $[\alpha^*]B \in \Gamma$ and $\mathcal{M}^P \models_s [\alpha^*]B$, then $\mathcal{M}^P \models_t [\alpha^*]B$,

by an argument just like that in 9.8, using $\mathcal{M}^P \models [\alpha^*]B \to [\alpha][\alpha^*]B$ (from *Mix*). Thus if $|s|R_{\alpha^*}|t|$, then $|s|R^n_\alpha|t|$ for some n, so that if $\mathcal{M}^P \models_s [\alpha^*]B$, we get $\mathcal{M}^P \models_t [\alpha^*]B$, and so $\mathcal{M}^P \models_t B$ as $\mathcal{M}^P \models [\alpha^*]B \to B$ by *Mix* again.

Filtration Lemma 10.8. *For any $B \in \Gamma$,*

$$\mathcal{M}^P \models_s B \quad \text{iff} \quad \mathcal{M}_\Gamma \models_{|s|} B.$$

Proof. From 10.7, in the usual way.

Corollary 10.9. *$\mathcal{M}_\Gamma$ is a standard model.*

Proof. The Filtration Lemma, and the definition of $R_{B?}$, give

$$R_{B?} = \{(x,x) : \mathcal{M}_\Gamma \models_x B\}$$

for $B? \in Prog_\Gamma$, which was the only standard-model condition not already guaranteed by the definition of $\mathcal{M}_\Gamma$.

The final step in the argument that *PDL* has the finite model property with respect to standard models, and is decidable, should by now be familiar to the reader.

Exercises 10.10

(1) Extend the syntax to include programs of the form α^{-1}, with the semantics

$$R_{\alpha^{-1}} = \{(t,s) : sR_\alpha t\}.$$

(2) Adapt the syntax to take the construction "**while** A **do** α" as primitive instead of α^*. Define standard models appropriately, and show that the resulting logic is axiomatised by replacing *Mix* and *Ind* by the schemata

$$A \to ([\textbf{while } A \textbf{ do } \alpha]B \to [\alpha][\textbf{while } A \textbf{ do } \alpha]B),$$
$$\neg A \to \langle\textbf{while } A \textbf{ do } \alpha\rangle\top,$$

and the well known *Iteration Rule* of Hoare:

$$\begin{array}{ll} \text{from} & \vdash A \wedge B \to [\alpha]B \\ \text{infer} & \vdash B \to [\textbf{while } A \textbf{ do } \alpha](B \wedge \neg A) \end{array}$$

(cf. Goldblatt [1982i]).

Concurrent Dynamic Logic

We now consider an extension of *PDL*, due to Peleg [1987], which introduces the *combination* $\alpha \cap \beta$ of commands α and β, interpreted as "α and β executed in parallel". Thus, whereas the theory of §9 envisaged a collection of processes taking turns to act, here we imagine processes acting independently *at the same time*. For example, we might contemplate a command of the form ***goto l and m***, which causes a program to execute the commands labelled by ***l*** and ***m*** simultaneously and in parallel.

In this context, the result of an execution started in state s will not be a single terminal state t, but rather a set T of states representing the terminal situations of all the parallel processes involved. Thus the relation R_α interpreting command α is no longer a set of pairs (s,t), but rather a set of pairs (s,T), with s a member of the state-set S, and $T \subseteq S$. So instead of $R_\alpha \subseteq S \times S$, we have $R_\alpha \subseteq S \times 2^S$.

To keep the two types of relation distinct, we will refer to a subset of $S \times S$ simply as a *binary relation* on S, and a subset of $S \times 2^S$ as a *reachability relation* on S. When $sR_\alpha T$, this signifies that T is "reachable" from s by an execution of α. There may be many ways of executing α, and hence many different state-sets T reachable from s by doing α.

To model the meaning of $\langle\alpha\rangle A$ as "there is an execution of α that terminates with A true", we specify

$$\mathcal{M} \models_s \langle\alpha\rangle A \quad \text{iff} \quad \text{there exists } T \subseteq S \text{ with } sR_\alpha T \text{ and } T \subseteq \mathcal{M}(A), \quad \text{(i)}$$

where

$$\mathcal{M}(A) = \{t \in S : \mathcal{M} \models_t A\}.$$

If $[\alpha]$ is identified with $\neg{<}\alpha{>}\neg$, as in Peleg [1987], the condition for truth of $[\alpha]A$ at s becomes

$$sR_\alpha T \quad \text{implies} \quad T \cap \mathcal{M}(A) \neq \emptyset.$$

Nerode and Wijesekera [1990] suggest that in this context a more appropriate modelling of "after every terminating execution of α, A is true", would be

$$\mathcal{M} \models_s [\alpha]A \quad \text{iff} \quad sR_\alpha T \quad \text{implies} \quad T \subseteq \mathcal{M}(A), \tag{ii}$$

making $[\alpha]$ and ${<}\alpha{>}$ no longer interdefinable via $\neg$.

The extension of PDL with $[\alpha]$ and ${<}\alpha{>}$ interpreted according to (i) and (ii) has not been investigated in the literature to date. Here we will demonstrate finite axiomatisability and decidability for this extension, by developing a new theory of canonical models and filtrations for reachability relations.

Notice that if a binary relation $\overline{R_\alpha}$ is defined by

$$s\overline{R_\alpha}t \quad \text{iff} \quad t \in \bigcup\{T : sR_\alpha T\},$$

then (ii) becomes

$$\mathcal{M} \models_s [\alpha]A \quad \text{iff} \quad s\overline{R_\alpha}t \quad \text{implies} \quad \mathcal{M} \models_t A.$$

This observation will allow us to relate much of the new theory of $[\alpha]$ given by (ii) to our earlier analysis of the binary relation semantics for PDL. At the same time, a whole new analysis is needed for ${<}\alpha{>}$.

Syntax and Semantics

The formal language of *Concurrent Propositional Dynamic Logic* ($CPDL$) is as for PDL, with the addition of $\cap$ and the independent treatment of $[\alpha]$ and ${<}\alpha{>}$:

Atomic formulae:	$p \in \Phi$
Atomic programs:	$\pi \in \Pi$
Formulae:	$A \in Fma(\Phi, \Pi)$
Programs:	$\alpha \in Prog(\Phi, \Pi)$

$$A ::= p \mid \bot \mid A_1 \to A_2 \mid {<}\alpha{>}A \mid [\alpha]A$$

$$\alpha ::= \pi \mid \alpha_1;\alpha_2 \mid \alpha_1 \cup \alpha_2 \mid \alpha_1 \cap \alpha_2 \mid \alpha^* \mid A?$$

A $CPDL$-model is a structure

$$\mathcal{M} = (S, \{R_\alpha : \alpha \in Prog(\Phi, \Pi)\}, V),$$

with R_α a reachability relation on S for each program α, i.e. $R_\alpha \subseteq S \times 2^S$, and the truth relation $\mathcal{M} \models_s A$ determined by (i) and (ii) above.

Operations on Reachability Relations

Let R and Q be reachability relations on a set S.

Composition. The relation $R \cdot Q \subseteq S \times 2^S$ is defined by

$s(R \cdot Q)T$ iff there exist $U \subseteq S$ with sRU, and a collection $\{T_u : u \in U\}$ of subsets of T with uQT_u for all $u \in U$, such that $T = \bigcup\{T_u : u \in U\}$.

Combination.

$$R \otimes Q = \{(s, T \cup W) : sRT \text{ and } sQW\}.$$

Iteration. Let

$$Id = \{(s, \{s\}) : s \in S\},$$

and define a sequence of reachability relations $R^{(n)}$ inductively by

$$\begin{aligned} R^{(0)} &= Id \\ R^{(n+1)} &= Id \cup R \cdot R^{(n)}. \end{aligned}$$

Then put

$$R^{(*)} = \bigcup\{R^{(n)} : n \in \omega\}.$$

Exercises 10.11

(1) $Q \subseteq Q'$ implies $R \cdot Q \subseteq R \cdot Q'$.

(2) $(R \cup R') \cdot Q = R \cdot Q \cup R' \cdot Q$.

(3) Give a counter-example to the assertion

$$R \cdot (Q \cup Q') = R \cdot Q \cup R \cdot Q'.$$

(4) $R^{(n)} \subseteq R^{(n+1)}$. Hence the operation $R^{(n)}$ is monotonic in n: $n \leq m$ implies $R^{(n)} \subseteq R^{(m)}$.

Standard Models

A $CPDL$-model is *standard* if it satisfies

$$\begin{aligned} R_{\alpha;\beta} &= R_\alpha \cdot R_\beta; \\ R_{\alpha\cup\beta} &= R_\alpha \cup R_\beta; \\ R_{\alpha\cap\beta} &= R_\alpha \otimes R_\beta; \\ R_{\alpha^*} &= R_\alpha^{(*)}; \\ R_{A?} &= \{(s, \{s\}) : \mathcal{M} \models_s A\}. \end{aligned}$$

Thus in a standard model, $R_{\mathbf{skip}} = Id$. The standard-model condition on $\cap$ ensures that $<\alpha \cap \beta>A$ gets the meaning "α and β can be executed in parallel so that on termination (in both computations) A is true".

To understand the meaning of the new iteration operation $R_\alpha^{(*)}$ that interprets α^*, consider the schema

$$<\alpha^*>A \leftrightarrow A \vee <\alpha><\alpha^*>A, \tag{iii}$$

which intuitively is true under the intended meaning of α^* as "repeat α some finite number (≥ 0) of times". In the binary relation semantics for PDL, where R_{α^*} is the ancestral R_α^*, truth of this schema in standard models is a consequence of the fact that

$$R_\alpha^* = id \cup R_\alpha \circ R_\alpha^*,$$

where

$$id = \{(s, s) : s \in S\}.$$

(Note also that in such standard models, $id = R_{\mathbf{skip}}$, and $A \leftrightarrow <\mathbf{skip}>A$ is true.)

Now in fact to have (iii) come out true in a PDL-model, it suffices to interpret α^* by any binary relation Q satisfying

$$Q = id \cup R_\alpha \circ Q. \tag{iv}$$

The characteristic property of the ancestral R_α^* is that it is the *least* solution of equation (iv), i.e. if (iv) holds then $R_\alpha^* \subseteq Q$ (cf. Exercise 1.5(4)). Thus in a model in which (iii) is true, we must have $R_\alpha^* \subseteq R_{\alpha^*}$. But then by requiring R_{α^*} itself to be the least solution of (iv) we add the converse inclusion $R_{\alpha^*} \subseteq R_\alpha^*$, which is just what is necessary to verify the PDL-axiom Ind.

Now if we put

$$F(Q) = id \cup R_\alpha \circ Q$$

for an arbitrary binary relation Q, then (iv) asserts that Q is a *fixed point* of the operator F, i.e. $F(Q) = Q$. There is a general theory about fixed points of operators like F that is fundamental to the study of recursive definitions: putting $F^{(0)} = F(\emptyset)$, and $F^{(n+1)} = F(F^{(n)})$, then knowing only that F is *monotonic*, i.e. that

$$Q \subseteq Q' \quad \text{implies} \quad F(Q) \subseteq F(Q'),$$

it can be shown that F must have a least fixed point, namely the relation

$$\bigcup\{F^{(n)} : n \in \omega\}.$$

We applied this theory above in defining $R^{(*)}$, using the monotonic operator

$$F(Q) = Id \cup R \cdot Q$$

on reachability relations Q (cf. Exercise 10.11(1)). Thus $R^{(*)}$ is defined as the least solution of the equation

$$Q = Id \cup R \cdot Q,$$

and so $R_\alpha^{(*)}$ in turn is the least reachability relation that interprets α^* to make schema (iii) come out true.

Further insight into the nature of the relation $R^{(*)}$ is given in Theorem 10.14(7) below.

Exercises 10.12

Define programs $\alpha^{(n)}$ by

$$\alpha^{(0)} = \mathbf{skip}$$
$$\alpha^{(n+1)} = \mathbf{skip} \cup (\alpha; \alpha^{(n)})$$

Prove the following in any *standard* model.

(1) $R_{\alpha^{(n)}} = R_\alpha^{(n)}$.

(2) $\mathcal{M} \models_s [\alpha^{(n)}]A$ iff $sR_\alpha^{(n)}T$ implies $T \subseteq \mathcal{M}(A)$

iff $s\overline{R_\alpha^{(n)}}t$ implies $\mathcal{M} \models_t A$ (cf. 10.13 below).

(3) $\mathcal{M} \models_s <\alpha^{(n)}>A$ iff there exists T with $sR_\alpha^{(n)}T$ and $T \subseteq \mathcal{M}(A)$.

(4) $\mathcal{M} \models_s [\alpha^*]A$ iff for all $n \geq 0$, $\mathcal{M} \models_s [\alpha^{(n)}]A$.

(5) $\mathcal{M} \models_s <\alpha^*>A$ iff for some $n \geq 0$, $\mathcal{M} \models_s <\alpha^{(n)}>A$.

Reduction to Binary Relations

For an arbitrary reachability relation R, define the binary relation $\overline{R}$ by

$$s\overline{R}t \quad \text{iff} \quad t \in \bigcup\{T : sRT\}$$
$$\text{iff} \quad \text{for some } T \subseteq S,\ sRT \text{ and } t \in T.$$

Lemma 10.13. *For any CPDL-model $\mathcal{M}$, standard or not,*

$$\mathcal{M} \models_s [\alpha]A \quad \textit{iff} \quad s\overline{R_\alpha}t \quad \textit{implies} \quad \mathcal{M} \models_t A.$$

We now investigate the properties of the relation $\overline{R}$, and for this purpose we need the binary relations $\overline{R}^n$, defined as on page 9. These satisfy

$$\overline{R}^0 = id$$
$$\overline{R}^{n+1} = \overline{R} \circ \overline{R}^n = \overline{R}^n \circ \overline{R}$$
$$\overline{R}^* = \bigcup\{\overline{R}^n : n \in \omega\}.$$

Theorem 10.14. *For any reachability relations* R_i, R, Q*:*

(1) $\overline{\bigcup_{i\in I} R_i} = \bigcup_{i\in I} \overline{R_i}$.

(2) $R \subseteq Q$ *implies* $\overline{R} \subseteq \overline{Q}$.

(3) $\overline{R \cdot Q} \subseteq \overline{R} \circ \overline{Q}$.

(4) *If* $Id \subseteq Q$, *then* $\overline{R \cdot Q} = \overline{R} \circ \overline{Q}$.

(5) $\overline{R^{(n+1)}} = id \cup \overline{R} \circ \overline{R^{(n)}}$.

(6) $\overline{R^{(n)}} = \overline{R}^0 \cup \cdots \cup \overline{R}^n$.

(7) $\overline{R^{(*)}} = \overline{R}^*$.

Proof. (1) and (2) are straightforward, and left as exercises.

(3) Suppose that $s\overline{R \cdot Q}t$. Then $s(R \cdot Q)T$ for some T with $t \in T$. From the definition of $R \cdot Q$, it follows that there exists U with sRU, and some $u \in U$ for which there is a $T_u \subseteq T$ with uQT_u and $t \in T_u$. But then $s\overline{R}u$ and $u\overline{Q}t$, showing that $s\overline{R} \circ \overline{Q}t$.

(4) If $Id \subseteq Q$, we want the converse of (3). Suppose then that $s\overline{R} \circ \overline{Q}t$, so that $s\overline{R}u$ and $u\overline{Q}t$ for some u. Then sRU for some U with $u \in U$, and uQT_u for some T_u with $t \in T_u$. Let

$$T = \bigcup \{\{v\} : u \neq v \in U\} \cup T_u.$$

Since $Id \subseteq Q$, we have $vQ\{v\}$ in general, so it follows (with $T_v = \{v\}$ for $v \neq u$) that $s(R \cdot Q)T$, and hence as $t \in T$ that $s\overline{R \cdot Q}t$.

(5) Since $Id \subseteq R^{(n)}$, $\overline{R \cdot R^{(n)}} = \overline{R} \circ \overline{R^{(n)}}$ by (4). But as $\overline{Id} = id$, (5) then follows from the definition of $R^{(n+1)}$ and (1).

(6) By induction on n. The case $n = 0$ asserts that $\overline{R^{(0)}} = \overline{R}^0$, which is just the true statement that $\overline{Id} = id$.
Assuming the result for n, from (5) and this induction hypothesis we then get

$$\begin{aligned}\overline{R^{(n+1)}} &= id \cup \overline{R} \circ (\overline{R}^0 \cup \cdots \cup \overline{R}^n)\\ &= \overline{R}^0 \cup (\overline{R} \circ \overline{R}^0 \cup \cdots \cup \overline{R} \circ \overline{R}^n)\\ &= \overline{R}^0 \cup \overline{R}^1 \cup \cdots \cup \overline{R}^{n+1},\end{aligned}$$

which gives the result for $n + 1$.

(7) From the definition of $\overline{R^{(*)}}$, applying (1) and then (5), we calculate

$$\begin{aligned}\overline{R^{(*)}} &= \overline{\bigcup_{n\in\omega} R^{(n)}}\\ &= \bigcup_{n\in\omega} \overline{R^{(n)}}\\ &= \bigcup_{n\in\omega} (\overline{R}^0 \cup \cdots \cup \overline{R}^n)\\ &= \bigcup_{n\in\omega} \overline{R}^n\\ &= \overline{R}^*\end{aligned}$$

Corollary 10.15. *In a standard model* $\mathcal{M}$,

$$\mathcal{M} \models_s [\alpha^*]A \quad \textit{iff} \quad s\overline{R_\alpha}^* t \quad \textit{implies} \quad \mathcal{M} \models_t A.$$

Proof. In a standard model, 10.14(7) implies $\overline{R_{\alpha^*}} = \overline{R_\alpha}^*$, so the result follows from Lemma 10.13.

This Corollary simplifies the determination of truth-values of formulae containing $[\alpha^*]$. For instance, it makes it easy to show that the PDL-axiom Ind is true in standard $CPLD$-models.

Exercises 10.16

Let $\mathcal{M}$ be standard.

(1) Prove by induction on n that

$$\mathcal{M} \models [\alpha^*](<\alpha>A \to A) \to [\alpha^*](<\alpha^{(n)}>A \to A).$$

(2) Use (1) and 10.12(5) to deduce that

$$\mathcal{M} \models [\alpha^*](<\alpha>A \to A) \to (<\alpha^*>A \to A).$$

Axioms for $CPDL$

Let $CPDL$ be the smallest logic in $Fma(\Phi, \Pi)$ that contains the schemata

B-K:	$[\alpha](A \to B) \to ([\alpha]A \to [\alpha]B)$,
B-$Comp$:	$[\alpha;\beta]A \leftrightarrow [\alpha][\beta]A$,
B-Alt:	$[\alpha \cup \beta]A \leftrightarrow [\alpha]A \wedge [\beta]A$,
B-$Comb$:	$[\alpha \cap \beta]A \leftrightarrow (<\alpha>\top \to [\beta]A) \wedge (<\beta>\top \to [\alpha]A)$,
B-Mix:	$[\alpha^*]A \to A \wedge [\alpha][\alpha^*]A$,
B-Ind:	$[\alpha^*](A \to [\alpha]A) \to (A \to [\alpha^*]A)$,
B-$Test$:	$[A?]B \leftrightarrow (A \to B)$,
D-K:	$[\alpha](A \to B) \to (<\alpha>A \to <\alpha>B)$,
D-$Comp$:	$<\alpha;\beta>A \leftrightarrow <\alpha><\beta>A$,
D-Alt:	$<\alpha \cup \beta>A \leftrightarrow <\alpha>A \vee <\beta>A$,
D-$Comb$:	$<\alpha \cap \beta>A \leftrightarrow <\alpha>A \wedge <\beta>A$,
D-Mix:	$A \vee <\alpha><\alpha^*>A \to <\alpha^*>A$,
D-Ind:	$[\alpha^*](<\alpha>A \to A) \to (<\alpha^*>A \to A)$,
D-$Test$:	$<A?>B \leftrightarrow (A \wedge B)$,
B-D:	$[\alpha]\bot \vee <\alpha>\top$,

and is closed under Necessitation for $[\alpha]$. Thus $CPDL$ is a normal logic. (The B- and D- prefixes stand for "Box" and "Diamond".) For the sake of legibility we will abbreviate $\vdash_{CPDL} A$ simply to $\vdash A$.

It will be shown that this logic has the finite model property with respect to standard $CPDL$-models.

Exercises 10.17

(1) (*Soundness*) If $\vdash A$, then A is true in all standard $CPDL$-models.

(2) $\vdash A \to B$ implies $\vdash [\alpha]A \to [\alpha]B$.

(3) $\vdash A \to B$ implies $\vdash <\alpha>A \to <\alpha>B$.

(4) $\vdash [\alpha]A \vee <\alpha>\top$.

(5) $\vdash [\alpha]A \to (<\alpha>B \to <\alpha>(A \wedge B))$.

Maximal Sets

Let S^m be the set of all $CPDL$-maximal subsets of $Fma(\Phi, \Pi)$. For each formula A, let

$$\|A\| = \{s \in S^m : A \in s\}.$$

For each $s \in S^m$ and program α, let

$$s_\alpha = \{A : [\alpha]A \in s\}, \qquad \text{and}$$

$$\|s_\alpha\| = \{t \in S^m : s_\alpha \subseteq t\}.$$

Thus $\|s_\alpha\| = \bigcap\{\|A\| : [\alpha]A \in s\}$.

Note that the condition "$s_\alpha \subseteq t$" is equivalent to "$sR^P_\alpha t$", which defines the binary relations in the canonical model for PDL.

Theorem 10.18.

(1) $\vdash A$ *iff* $\|A\| = S^m$.

(2) $\vdash A \to B$ *iff* $\|A\| \subseteq \|B\|$.

(3) $\|A \vee B\| = \|A\| \cup \|B\|$.

(4) $\|A \wedge B\| = \|A\| \cap \|B\|$.

(5) $\|s_\alpha\| \subseteq \|A\|$ *implies* $[\alpha]A \in s$.

(6) *If* $\|s_\alpha\| \cap \|B\| \subseteq \|A\|$ *and* $<\alpha>B \in s$, *then* $<\alpha>A \in s$.

(7) *If* $s, u \in S^m$ *and* $s_\alpha \subseteq u$, *then* $\|u_\beta\| \subseteq \|s_{\alpha;\beta}\|$.

(8) $\|s_{\alpha\cup\beta}\| = \|s_\alpha\| \cup \|s_\beta\|$.

(9) *If* $<\alpha>\top, <\beta>\top \in s$, *then* $\|s_{\alpha\cap\beta}\| = \|s_\alpha\| \cup \|s_\beta\|$.

Proof. (1)–(4) are now familiar properties of maximal sets.

(5) This is essentially as in Theorem 3.2. If $\|s_\alpha\| \subseteq \|A\|$, then every maximal extension of s_α contains A, and so by 2.6(1), $s_\alpha \vdash A$. Hence

$$\vdash B_0 \to (B_1 \to (\cdots \to (B_{n-1} \to A)\cdots))$$

for some n, and some formulae B_i with $[\alpha]B_i \in s$. Then using Necessitation (directly if $n = 0$) and axiom B-K,

$$\vdash [\alpha]B_0 \to ([\alpha]B_1 \to (\cdots \to ([\alpha]B_{n-1} \to [\alpha]A)\cdots)),$$

from which $[\alpha]A \in s$ follows because s contains all theorems and is closed under Detachment.

(6) Let $t \in S$ have $s_\alpha \subseteq t$. Then if $B \in t$, $t \in \|s_\alpha\| \cap \|B\|$, so as $\|s_\alpha\| \cap \|B\| \subseteq \|A\|$, then $A \in t$. Thus $(B \to A) \in t$. This shows that $\|s_\alpha\| \subseteq \|B \to A\|$, so by (5), $[\alpha](B \to A) \in s$. But then by axiom D-K, $(<\alpha>B \to <\alpha>A) \in s$, giving the desired result that if $<\alpha>B \in s$ then $<\alpha>A \in s$.

(7) Let $s_\alpha \subseteq u$. Then if $t \in \|u_\beta\|$, we reason as follows. If $A \in s_{\alpha;\beta}$, then $[\alpha;\beta]A \in s$, so $[\alpha][\beta]A \in s$ by axiom B-*Comp*, whence $[\beta]A \in s_\alpha \subseteq u$, giving $A \in u_\beta \subseteq t$. This shows $s_{\alpha;\beta} \subseteq t$, i.e. $t \in \|s_{\alpha;\beta}\|$.

(8) Here we want to show that

$$s_{\alpha\cup\beta} \subseteq t \quad \text{iff} \quad s_\alpha \subseteq t \text{ or } s_\beta \subseteq t.$$

The implication from right to left is straightforward, with the aid of B-*Alt*. For the converse, suppose that $s_\alpha \not\subseteq t$ and $s_\beta \not\subseteq t$. Then there must be formulae A and B with $[\alpha]A, [\beta]B \in s$, but $A \notin t$ and $B \notin t$. Now $[\alpha]A \to [\alpha](A \vee B)$ is a theorem (cf. 10.17(2)), so $[\alpha](A \vee B) \in s$. Similarly, $[\beta](A \vee B) \in s$. Hence by B-*Alt*, $[\alpha \cup \beta](A \vee B) \in s$. Since $(A \vee B) \notin s$, this shows that $s_{\alpha\cup\beta} \not\subseteq t$.

(9) If $<\alpha>\top, <\beta>\top \in s$, then by axiom B-*Comb*,

$$[\alpha \cap \beta]A \in s \quad \text{iff} \quad [\alpha]A \in s \text{ and } [\beta]A \in s.$$

But this allows us to prove that

$$s_{\alpha\cap\beta} \subseteq t \quad \text{iff} \quad s_\alpha \subseteq t \text{ or } s_\beta \subseteq t,$$

in the same manner as for (8).

Reachability for Maximal Sets

Let $s \in S^m$ and $T \subseteq S^m$. For each program α, put

$$sR_\alpha T \quad \text{iff} \quad \text{there exists } B \text{ with } <\alpha>B \in s \text{ and } T = \|s_\alpha\| \cap \|B\|.$$

Theorem 10.19.

(1) $<\alpha>A \in s$ *iff there exists* T *with* $sR_\alpha T$ *and* $T \subseteq \|A\|$.

(2) $<\alpha>\top \in s$ *implies* $sR_\alpha\|s_\alpha\|$.

(3) $s\overline{R_\alpha}t$ *iff* $s_\alpha \subseteq t$.

(4) $[\alpha]A \in s$ *iff* $sR_\alpha T$ *implies* $T \subseteq \|A\|$.

Proof.

(1) If $<\alpha>A \in s$, then defining $T = \|s_\alpha\| \cap \|A\|$ immediately gives $sR_\alpha T$ and $T \subseteq \|A\|$. Conversely, if $sR_\alpha T \subseteq \|A\|$, then there exists B with $<\alpha>B \in s$ and $T = \|s_\alpha\| \cap \|B\|$. But then $\|s_\alpha\| \cap \|B\| \subseteq \|A\|$, so Theorem 10.18(6) gives $<\alpha>A \in s$, as desired.

(2) From the definition of R_α, since $\|s_\alpha\| \cap \|\top\| = \|s_\alpha\|$.

(3) If $s\overline{R_\alpha}t$, then $t \in T$ for some T of the form $\|s_\alpha\| \cap \|B\|$. But then $t \in \|s_\alpha\|$, i.e. $s_\alpha \subseteq t$.
Conversely, if $s_\alpha \subseteq t$, then since $\bot \notin t$, we get $[\alpha]\bot \notin s$, so by axiom B-D, $<\alpha>\top \in s$. Hence by (2), $sR_\alpha\|s_\alpha\|$. Since $t \in \|s_\alpha\|$, this gives $s\overline{R_\alpha}t$.

(4) By Theorem 10.18(5) and the definition of s_α, it follows that to have $[\alpha]A \in s$ it is necessary and sufficient that

$$s_\alpha \subseteq t \quad \text{implies} \quad A \in t,$$

which is equivalent by (3) to

$$s\overline{R_\alpha}t \quad \text{implies} \quad A \in t,$$

which in turn holds if, and only if,

$$sR_\alpha T \quad \text{implies} \quad T \subseteq \|A\|.$$

Corollary 10.20. *If there exists some t with $s\overline{R_\alpha}t$, then $<\alpha>\top \in s$.*

Proof. If $s\overline{R_\alpha}t$, there must be some T with $sR_\alpha T$. Since $T \subseteq \|\top\|$, 10.19(1) then gives $<\alpha>\top \in s$.

Canonical Model

The canonical model for $CPDL$ is the structure

$$\mathcal{M}^m = (S^m, \{R_\alpha : \alpha \in Prog(\Phi, \Pi)\}, V^m),$$

where S^m is the set of all $CPDL$-maximal sets, R_α is as defined prior to Theorem 10.19, and $V^m(p) = \{s \in S^m : p \in s\}$ as usual.

Note that in this model the relation $\overline{R_\alpha}$ is identical to R^P_α, by 10.19(3).

Truth Lemma 10.21. *For any* $A \in Fma(\Phi, \Pi)$,

$$\mathcal{M}^m(A) = \|A\|,$$

i.e. for all $s \in S^m$,

$$\mathcal{M}^m \models_s A \quad \text{iff} \quad A \in s.$$

Proof. By induction on the formation of A in the usual way, with the key inductive cases for $<\alpha>$ and $[\alpha]$ provided by 10.19(1) and 10.19(4), respectively.

As with PDL, the canonical model $\mathcal{M}^m$ determines the logic $CPDL$, but cannot be shown to be standard. Some properties that it does enjoy, and that will be used in our completeness theorem, are collected in the next result.

Theorem 10.22. *The following hold in the canonical* $CPDL$*-model.*

(1) *Tests are standard, i.e.* $sR_{A?}T$ *iff* $T = \{s\}$ *and* $\mathcal{M}^m \models_s A$.

(2) *If* $sR_{\alpha;\beta}T$, *then* $s(R_\alpha \cdot R_\beta)W$ *for some* $W \subseteq T$.

(3) *If* $sR_{\alpha\cup\beta}T$, *then* $s(R_\alpha \cup R_\beta)W$ *for some* $W \subseteq T$.

(4) $R_{\alpha\cap\beta} \subseteq R_\alpha \otimes R_\beta$.

Proof.

(1) Noting that $\mathcal{M}^m \models_s A$ iff $A \in s$, we have that if $\mathcal{M}^m \models_s A$, then $B \in s$ iff $(A \to B) \in s$ for any formula B, so by axiom B-*Test*, $[A?]B \in s$ iff $B \in s$, showing that $s_{A?} = s$. Moreover, this in turn implies that $\|s_{A?}\| = \{s\}$, since s is maximal.

Thus if $sR_{A?}T$, then $T = \|s_{A?}\| \cap \|B\|$ for some B with $<A?>B \in s$. Hence from axiom D-*Test*, $A, B \in s$, whence $\|s_{A?}\| = \{s\}$ as above, and $\{s\} \subseteq \|B\|$. Thus $T = \{s\} \cap \|B\| = \{s\}$, with $\mathcal{M}^m \models_s A$ as desired.

Conversely, if $\mathcal{M}^m \models_s A$ and $T = \{s\}$, then $\|s_{A?}\| = \{s\}$ and $T = \|s_{A?}\| \cap \|A\|$. Hence $sR_{A?}T$, since D-*Test* gives $<A?>A \in s$.

(2) Let $sR_{\alpha;\beta}T$. Then $T = \|s_{\alpha;\beta}\| \cap \|A\|$ for some A with $<\alpha;\beta>A \in s$. Then by D-*Comp*, $<\alpha><\beta>A \in s$, so $sR_\alpha U$, where $U = \|s_\alpha\| \cap \|<\beta>A\|$.

For each $u \in U$, put $T_u = \|u_\beta\| \cap \|A\|$, so that $uR_\beta T_u$, since $<\beta>A \in u$. Also, as $u \in \|s_\alpha\|$, i.e. $s_\alpha \subseteq u$, Theorem 10.18(7) yields $\|u_\beta\| \subseteq \|s_{\alpha;\beta}\|$, showing that $T_u \subseteq T$. Thus the desired result follows by putting $W = \bigcup\{T_u : u \in U\}$.

(3) If $sR_{\alpha\cup\beta}T$, then $T = \|s_{\alpha\cup\beta}\| \cap \|A\|$ for some A with $<\alpha\cup\beta>A \in s$. Axiom D-*Alt* then implies that one of $<\alpha>A$ and $<\beta>A$ is in s. If, say, $<\alpha>A \in s$, then $sR_\alpha W$, where $W = \|s_\alpha\| \cap \|A\|$. By Theorem

10.18(8), $\|s_\alpha\| \subseteq \|s_{\alpha\cup\beta}\|$, so $W \subseteq T$. Similarly, if $<\beta>A \in s$, we take $W = \|s_\beta\| \cap \|B\|$, and get $sR_\beta W \subseteq T$. In either case, $s(R_\alpha \cup R_\beta)W \subseteq T$.

(4) If $sR_{\alpha\cap\beta}T$, then $T = \|s_{\alpha\cap\beta}\| \cap \|A\|$ for some A with $<\alpha \cap \beta>A \in s$. Then by D-*Comb*, $<\alpha>A, <\beta>A \in s$, so $sR_\alpha(\|s_\alpha\| \cap \|A\|)$ and $sR_\beta(\|s_\beta\| \cap \|A\|)$. Hence $s(R_\alpha \otimes R_\beta)U$, where

$$U = (\|s_\alpha\| \cap \|A\|) \cup (\|s_\beta\| \cap \|A\|) = (\|s_\alpha\| \cup \|s_\beta\|) \cap \|A\|.$$

Since $\vdash <\alpha>A \to <\alpha>\top$ (10.17(3)) and $<\alpha>A \in s$, it follows that $<\alpha>\top \in s$. Similarly $<\beta>\top \in s$. But then by 10.18(9) $U = T$.

Execution Relations

If $s\overline{R_\alpha}t$, then intuitively there is an execution of α from s that produces a set T of terminal states including t. We may regard this execution as generating a *tree* of states, with T being the set of leaves of the tree. There will be a path through this tree from s to t, comprising a sequence of executions of atomic programs and/or tests (cf. §2.2 of Peleg [1987i] for an indication of how to formalise this idea).

If further $t\overline{R_\beta}u$, then there will be a similar computation tree containing a path from t to u as a result of executing β from t. We then have $s\overline{R_\alpha} \circ \overline{R_\beta}u$, but we cannot conclude that $s\overline{R_{\alpha;\beta}}t$ without first showing that β-computation trees can be attached to every state in T, and not just t. Nonetheless one might suggest that u has been arrived at from s by an instance of "doing α and then β".

These observations may provide some motivation for the following technical definition of relations R^+_α whose chief purpose is to give a representation of program composition $\alpha;\beta$ by binary relation composition $\circ$, and which will be used in defining filtrations of *CPDL*-models.

Given a *CPDL*-model

$$\mathcal{M} = (S, \{R_\alpha : \alpha \in Prog(\Phi, \Pi)\}, V),$$

define a family $\{R^+_\alpha : \alpha \in Prog(\Phi, \Pi)\}$ of binary relations on S inductively by

$$\begin{aligned} R^+_\pi &= \overline{R_\pi}\,; \\ R^+_{A?} &= \overline{R_{A?}}\,; \\ R^+_{\alpha;\beta} &= R^+_\alpha \circ R^+_\beta; \\ R^+_{\alpha\cup\beta} &= R^+_\alpha \cup R^+_\beta; \\ R^+_{\alpha^*} &= (R^+_\alpha)^*; \end{aligned}$$

and

$$sR^+_{\alpha\cap\beta}t \quad \text{iff} \quad \text{for some } T \text{, either}$$
$$\text{(i) } sR^+_\alpha t \text{ and } sR_\beta T, \quad \text{or}$$
$$\text{(ii) } sR_\alpha T \text{ and } sR^+_\beta t.$$

Theorem 10.23. *In a model that is standard except possibly for tests,* $\overline{R_\alpha} \subseteq R^+_\alpha$.

Proof. By induction on the formation of α. The cases $\alpha = \pi$ and $\alpha = A?$ are immediate by definition of R^+_α. For the inductive cases, assume the result for α and β.

Composition:

$$\begin{array}{lll} \overline{R_{\alpha;\beta}} & = \overline{R_\alpha \cdot R_\beta} & \text{standard condition for } \alpha;\beta \\ & \subseteq \overline{R_\alpha} \circ \overline{R_\beta} & 10.14(3) \\ & \subseteq R^+_\alpha \circ R^+_\beta & \text{hypothesis on } \alpha \text{ and } \beta \\ & = R^+_{\alpha;\beta}. & \end{array}$$

Alternation:

$$\begin{array}{lll} \overline{R_{\alpha\cup\beta}} & = \overline{R_\alpha \cup R_\beta} & \text{standard condition for } \alpha\cup\beta \\ & = \overline{R_\alpha} \cup \overline{R_\beta} & 10.14(1) \\ & \subseteq R^+_\alpha \cup R^+_\beta & \text{hypothesis on } \alpha \text{ and } \beta \\ & = R^+_{\alpha\cup\beta}. & \end{array}$$

Iteration:

$$\begin{array}{lll} \overline{R_{\alpha^*}} & = \overline{R^{(*)}_\alpha} & \text{standard condition for } \alpha^* \\ & = \overline{R_\alpha}^{\,*} & 10.14(7) \\ & \subseteq (R^+_\alpha)^* & \text{hypothesis on } \alpha \\ & = R^+_{\alpha^*}. & \end{array}$$

Combination: If $s\overline{R_{\alpha\cap\beta}}t$, then by the standard condition there are T, W with $sR_\alpha T$, $sR_\beta W$, and $t \in T \cup W$. Now if $t \in T$, then $s\overline{R_\alpha}t$, so $sR^+_\alpha t$ by the hypothesis on α, whence as $sR_\beta W$ we get $sR^+_{\alpha\cap\beta}t$. On the other hand, if $t \in W$ we similarly get $sR^+_\beta t$ and $sR_\alpha T$, leading again to the desired conclusion $sR^+_{\alpha\cap\beta}t$.

Theorem 10.24. *Let $\mathcal{M}$ be a model that is standard except possibly for tests. If α is any program, then for all formulae A we have*

$$\mathcal{M} \models_s [\alpha]A \quad \text{iff} \quad sR^+_\alpha t \quad \text{implies} \quad \mathcal{M} \models_t A.$$

Proof. Since in general

$$\mathcal{M} \models_s [\alpha]A \quad \text{iff} \quad s\overline{R_\alpha}t \quad \text{implies} \quad \mathcal{M} \models_t A$$

(Lemma 10.13), the fact that $\overline{R_\alpha} \subseteq R_\alpha^+$ implies directly that the statement of the Theorem holds from right to left. We prove the converse by induction on the formation of α.

The cases $\alpha = \pi$ and $\alpha = A?$ are immediate, as then $R_\alpha^+ = \overline{R_\alpha}$. For the inductive cases, assume the result for α and β.

Composition. Let $\mathcal{M} \models_s [\alpha;\beta]A$ and $sR^+_{\alpha;\beta}t$. Then there exists u with sR_α^+u and uR_β^+t. Since $\mathcal{M}$ is standard for composition, it verifies B-*Comp*, and so $\mathcal{M} \models_s [\alpha][\beta]A$. The induction hypothesis on α then gives $\mathcal{M} \models_u [\beta]A$, from which the hypothesis on β yields the desired conclusion $\mathcal{M} \models_t A$.

Alternation. If $\mathcal{M} \models_s [\alpha \cup \beta]A$ and $sR^+_{\alpha\cup\beta}t$, then either $sR^+_\alpha t$ or $sR^+_\beta t$, so as $\mathcal{M}$ verifies B-*Alt*, the hypothesis on α and β leads to $\mathcal{M} \models_t A$.

Iteration. Let $\mathcal{M} \models_s [\alpha^*]A$. Then we first show that for any n,

$$s(R^+_\alpha)^n t \quad \text{implies} \quad \mathcal{M} \models_t [\alpha^*]A. \tag{†}$$

The base case $n = 0$ is immediate, since then $s = t$. Assuming the result for n, suppose that $s(R^+_\alpha)^{n+1}t$. Then for some u, $s(R^+_\alpha)^n u$ and $uR^+_\alpha t$. By the hypothesis on n, $\mathcal{M} \models_u [\alpha^*]A$. Hence $\mathcal{M} \models_u [\alpha][\alpha^*]A$, since $\mathcal{M}$ verifies B-*Mix*, so by the hypothesis on α, $\mathcal{M} \models_t [\alpha^*]A$. This completes the inductive proof of (†).

Now if $sR^+_{\alpha^*}t$, then $s(R^+_\alpha)^n t$ for some n, and so $\mathcal{M} \models_t [\alpha^*]A$ by (†). Again since $\mathcal{M}$ verifies B-*Mix*, this implies $\mathcal{M} \models_t A$.

Combination. Let $\mathcal{M} \models_s [\alpha \cap \beta]A$ and $sR^+_{\alpha\cap\beta}t$. Then there exists T such that either (i) $sR^+_\alpha t$ and $sR_\beta T$, or else (ii) $sR_\alpha T$ and $sR^+_\beta t$.

Now if (i) holds, then $sR_\beta T$ implies $\mathcal{M} \models_s <\beta>\top$, so as $\mathcal{M}$ verifies B-*Comb*, $\mathcal{M} \models_s [\alpha]A$. But then the hypothesis on α gives $\mathcal{M} \models_t A$. Similarly, if (ii) holds we are led to $\mathcal{M} \models_t A$ by the other conjunct of B-*Comb* and the hypothesis on β.

Filtrations

To define filtrations of *CPDL*-models, a set Γ of formulae is defined to be *closed* if

Γ is closed under subformulae;

$[B?]D \in \Gamma$ implies $B \in \Gamma$;

$[\alpha;\beta]B \in \Gamma$ implies $[\alpha][\beta]B \in \Gamma$;

$[\alpha \cup \beta]B \in \Gamma$ implies $[\alpha]B, [\beta]B \in \Gamma$;

$[\alpha \cap \beta]B \in \Gamma$ implies $[\alpha]B, [\beta]B, <\alpha>\top, <\beta>\top \in \Gamma$;

$[\alpha^*]B \in \Gamma$ implies $[\alpha][\alpha^*]B \in \Gamma$;
$<B?>D \in \Gamma$ implies $B \in \Gamma$;
$<\alpha;\beta>B \in \Gamma$ implies $<\alpha><\beta>B \in \Gamma$;
$<\alpha \cup \beta>B \in \Gamma$ implies $<\alpha>B, <\beta>B \in \Gamma$;
$<\alpha \cap \beta>B \in \Gamma$ implies $<\alpha>B, <\beta>B \in \Gamma$;
$<\alpha^*>B \in \Gamma$ implies $<\alpha><\alpha^*>B \in \Gamma$.

By the same method as used in Lemma 10.5, it can be shown for the language of $CPDL$ that

Lemma 10.25. *For any $A \in Fma(\Phi, \Pi)$ there is a finite closed set Γ with $A \in \Gamma$.*

Now let Γ be a finite closed set. Put $\Phi_\Gamma = \Phi \cap \Gamma$, and let $Prog_\Gamma$ be the smallest set of programs that includes all atomic programs and tests occurring in members of Γ, and is closed under ;, $\cup$, $\cap$, and *. For $s, t \in S^m$, put

$$s \sim_\Gamma t \quad \text{iff} \quad s \cap \Gamma = t \cap \Gamma,$$
$$|s| = \{t \in S^m : s \sim_\Gamma t\},$$
$$S_\Gamma = \{|s| : s \in S^m\},$$

as usual, and for $T \subseteq S^m$, and $X \subseteq S_\Gamma$, put

$$|T| = \{|s| : s \in T\},$$
$$S_X = \{s \in S^m : |s| \in X\}.$$

Exercises 10.26

(1) $T \subseteq U$ implies $|T| \subseteq |U|$.
(2) $X \subseteq Y$ implies $S_X \subseteq S_Y$.
(3) $S_X \subseteq T$ implies $X \subseteq |T|$.
(4) $X = |S_X|$.
(5) $T \subseteq S_{|T|}$.
(6) $|s| = S_{\{|s|\}}$.

Now let

$$\mathcal{M} = (S_\Gamma, \{\rho_\alpha : \alpha \in Prog_\Gamma\}, V_\Gamma),$$

be a model based on S_Γ, with V_Γ the usual Φ_Γ-valuation. Then the reachability relation ρ_α on S_Γ is defined to be a *Γ-filtration* of the relation R_α from the canonical model $\mathcal{M}^m$ if, and only if, the following four conditions are satisfied.

(B1) $s\overline{R_\alpha}t$ implies $|s|\rho^+_\alpha|t|$.

(B2) $|s|\overline{\rho_\alpha}|t|$ implies $\{B : [\alpha]B \in s \cap \Gamma\} \subseteq t$.

(D1) $sR_\alpha T$ implies $|s|\rho_\alpha X$ for some $X \subseteq |T|$.

(D2) if $|s|\rho_\alpha X$ and $S_X \subseteq \|B\|$, then $<\alpha>B \in \Gamma$ implies $<\alpha>B \in s$.

ρ_α will be called *strong* if it satisfies

$$sR_\alpha T \text{ implies } |s|\rho_\alpha|T|.$$

Any strong relation ρ_α obviously satisfies (D1). But it also satisfies (B1) when $\overline{\rho_\alpha} \subseteq \rho^+_\alpha$, e.g. when $\mathcal{M}$ is standard except possibly for tests (10.23). For then if $s\overline{R_\alpha}t$, we have $sR_\alpha T$ for some T with $t \in T$, hence $|s|\rho_\alpha|T|$ and $|t| \in |T|$, showing $|s|\overline{\rho_\alpha}|t|$. But then $|s|\rho^+_\alpha|t|$ since $\overline{\rho_\alpha} \subseteq \rho^+_\alpha$.

The model $\mathcal{M}$ will be called a *Γ-filtration* of the canonical model $\mathcal{M}^m$ if ρ_α is a Γ-filtration of R_α for all $\alpha \in Prog_\Gamma$.

Filtration Lemma 10.27. *Let $\mathcal{M}$ be a Γ-filtration of $\mathcal{M}^m$ that is standard except possibly for tests. Then for any $B \in \Gamma$ and $s \in S^m$,*

$$\mathcal{M}^m \models_s B \quad \text{iff} \quad \mathcal{M} \models_{|s|} B.$$

Proof. By induction on the formation of B.

For the inductive case for $[\alpha]$, assume the result for B. Then if $[\alpha]B \in \Gamma$ and $\mathcal{M} \models_{|s|} [\alpha]B$, since $\mathcal{M}$ is standard except possibly for tests we get that

$$|s|\rho^+_\alpha|t| \quad \text{implies} \quad \mathcal{M} \models_{|t|} B,$$

by Theorem 10.24. From (B1) and the induction hypothesis on B, we then get

$$s\overline{R_\alpha}t \quad \text{implies} \quad \mathcal{M}^m \models_t B.$$

This in turn gives $\mathcal{M}^m \models_s [\alpha]B$ by Lemma 10.13.

Conversely, if $\mathcal{M}^m \models_s [\alpha]B$, i.e. $[\alpha]B \in s$, then from (B2) and the induction hypothesis we get that

$$|s|\overline{\rho_\alpha}|t| \quad \text{implies} \quad \mathcal{M} \models_{|t|} B,$$

which implies $\mathcal{M} \models_{|s|} [\alpha]B$ by 10.13 again.

Now for the inductive case of $<\alpha>$. First, if $<\alpha>B \in \Gamma$ and $\mathcal{M}^m \models_s <\alpha>B$, then there exists $T \subseteq S^m$ with $sR_\alpha T \subseteq \|B\|$. Thus if the Lemma holds for B, then for $t \in T$ we have $B \in t$, whence $\mathcal{M} \models_{|t|} B$, showing that $|T| \subseteq \mathcal{M}(B)$. But by (D1), $|s|\rho_\alpha X$ for some $X \subseteq |T|$. Then $X \subseteq \mathcal{M}(B)$, giving $\mathcal{M} \models_{|s|} <\alpha>B$.

Conversely, if $\mathcal{M} \models_{|s|} <\alpha>B$, then $|s|\rho_\alpha X$ for some $X \subseteq \mathcal{M}(B)$. The inductive hypothesis on B then yields $S_X \subseteq \|B\|$, and so (D2) gives $\mathcal{M}^m \models_s <\alpha>B$.

Existence of Filtrations

For $\alpha \in Prog_\Gamma$, define

$$
\begin{aligned}
|s|\rho^\lambda_\alpha X \quad \text{iff} \quad & \text{(i) } |t| \in X \quad \text{implies} \quad \{B : [\alpha]B \in s \cap \Gamma\} \subseteq t; \quad \text{and} \\
& \text{(ii) } S_X \subseteq \|B\| \text{ and } <\alpha>B \in \Gamma \quad \text{implies} \quad <\alpha>B \in s.
\end{aligned}
$$

Theorem 10.28. *If $\overline{\rho^\lambda_\alpha} \subseteq (\rho^\lambda_\alpha)^+$, then ρ^λ_α is a Γ-filtration of R_α, and is in fact the largest one.*

Proof. First we show that ρ^λ_α is strong, taking care of (B1) and (D1) since $\overline{\rho^\lambda_\alpha} \subseteq (\rho^\lambda_\alpha)^+$, as explained above. So, let $sR_\alpha T$, with the objective of showing that $|s|\rho^\lambda_\alpha|T|$, i.e. that (i) and (ii) above hold with $X = |T|$. We have $T = \|s_\alpha\| \cap \|C\|$, for some C with $<\alpha>C \in s$.

Now for (i), if $|t| \in |T|$, then $t \sim_\Gamma u$ for some $u \in T$, so that if $[\alpha]B \in s \cap \Gamma$ then $T \subseteq \|B\|$ as $sR_\alpha T$, hence $B \in u$, and so $B \in t$ as $B \in \Gamma$.

For (ii), suppose that $S_{|T|} \subseteq \|B\|$ and $<\alpha>B \in \Gamma$. Then as $T \subseteq S_{|T|}$, we have $sR_\alpha T \subseteq \|B\|$, and so $<\alpha>B \in s$ follows by Theorem 10.19(1). This completes the proof that ρ^λ_α is strong.

Next we show that (B2) holds for ρ^λ_α: if $|s|\overline{\rho^\lambda_\alpha}|t|$ then $|s|\rho^\lambda_\alpha X$ and $|t| \in X$ for some X, so that by part (i) of the definition of ρ^λ_α, $\{B : [\alpha]B \in s \cap \Gamma\} \subseteq t$.

Noting that (D2) for ρ^λ_α is immediate from (ii), we have now shown that ρ^λ_α is a filtration. The proof that it is the largest is left as an exercise.

The Finite Model

Given a finite closed Γ, construct a model

$$\mathcal{M}_\Gamma = (S_\Gamma, \{\rho_\alpha : \alpha \in Prog_\Gamma\}, V_\Gamma),$$

by letting ρ_π be *any* Γ-filtration of R_π (such existing by 10.28),

$$\rho_{B?} = \{(|s|, \{|s|\}) : \mathcal{M}^m \models_s B\},$$

and otherwise defining ρ_α inductively by the standard-model condition on α. *Thus $\mathcal{M}_\Gamma$ is standard except possibly for tests.*

Theorem 10.29. *$\mathcal{M}_\Gamma$ is a Γ-filtration of the canonical CPDL-model $\mathcal{M}^m$.*

Proof. We have to show that ρ_α is a Γ-filtration of R_α for each $\alpha \in Prog_\Gamma$.

Tests. Suppose $B? \in Prog_\Gamma$. If $sR_{B?}T$, then by 10.22(1), $T = \{s\}$ and $\mathcal{M}^m \models_s B$. Hence $|T| = \{|s|\}$, and so $|s|\rho_{B?}|T|$ by definition of $\rho_{B?}$. This shows that $\rho_{B?}$ is strong, and so fulfils (B1) and (D1).

For (B2), let $|s|\overline{\rho_{B?}}|t|$, so that $|s| = |t|$ and $B \in s$. Then if $[B?]D \in s \cap \Gamma$, we get $D \in s$ via B-*Test*, and so $D \in t$ as $s \sim_\Gamma t$.

For (D2), let $|s|\rho_{B?}X$ and $S_X \subseteq \|D\|$. Then $X = \{|s|\}$ and $B \in s$, so that $s \in S_X$, giving $D \in s$. Hence by D-*Test*, $<B?>D \in s$.

Composition. Suppose that $(\alpha;\beta) \in Prog_\Gamma$, and, inductively, that ρ_α and ρ_β are Γ-filtrations of R_α and R_β, respectively.

(B1): The argument is just as for PDL in Theorem 10.7. For $s \in S$, let A_s be a formula having

$$A_s \in t \quad \text{iff} \quad |s|\rho^+_{\alpha;\beta}|t|.$$

If $s\overline{R_\alpha}u\overline{R_\beta}t$, then by (B1) for α and β, $|s|\rho^+_\alpha|u|\rho^+_\beta|t|$. Hence $|s|\rho^+_\alpha \circ \rho^+_\beta|t|$, i.e. $|s|\rho^+_{\alpha;\beta}|t|$ by definition of $\rho^+_{\alpha;\beta}$, and so $A_s \in t$. This shows that $[\alpha][\beta]A_s \in s$, and hence by axiom B-*Comp*, $[\alpha;\beta]A_s \in s$ as needed to ensure that $s\overline{R_{\alpha;\beta}}t$ implies $|s|\rho^+_{\alpha;\beta}|t|$.

(B2): Let $|s|\overline{\rho_{\alpha;\beta}}|t|$, i.e. $|s|\overline{\rho_\alpha \cdot \rho_\beta}|t|$. Then $|s|\overline{\rho_\alpha} \circ \overline{\rho_\beta}|t|$ by 10.14(3), so for some u, $|s|\overline{\rho_\alpha}|u|$ and $|u|\overline{\rho_\beta}|t|$. Then if $[\alpha;\beta]B \in s \cap \Gamma$, $[\alpha][\beta]B \in s \cap \Gamma$ by *Comp*, so (B2) for α and β give $[\beta]B \in u$ and thence $B \in t$.

(D1): Let $sR_{\alpha;\beta}T$. Then by Theorem 10.22(2), there exists $U \subseteq S^m$ with $sR_\alpha U$, such that for each $u \in U$ there exists $T_u \subseteq T$ with $uR_\beta T_u$. By (D1) for α there exists $X \subseteq S_\Gamma$ with $|s|\rho_\alpha X \subseteq |U|$. Then if $x \in X$, we have $x = |u|$ for some $u \in U$, so by (D1) for β, there exists $Y_x \subseteq S_\Gamma$ with $x\rho_\beta Y_x \subseteq |T_u| \subseteq |T|$. Thus putting

$$Z = \bigcup\{Y_x : x \in X\},$$

we have $|s|(\rho_\alpha \cdot \rho_\beta)Z$, hence $|s|\rho_{\alpha;\beta}Z \subseteq |T|$.

(D2): If $|s|\rho_{\alpha;\beta}X$, i.e. $|s|(\rho_\alpha \cdot \rho_\beta)X$, then there exists $Y \subseteq S_\Gamma$ with $|s|\rho_\alpha Y$, such that $X = \bigcup\{X_y : y \in Y\}$, with $y\rho_\beta X_y$ for all $y \in Y$.

Now suppose $S_X \subseteq \|B\|$ and $<\alpha;\beta>B \in \Gamma$. We want $<\alpha;\beta>B \in s$. But if $t \in S_Y$, then $|t| \in Y$ and $S_{X_{|t|}} \subseteq S_X \subseteq \|B\|$, so as $<\beta>B \in \Gamma$ and $|t|\rho_\beta X_{|t|}$, (D2) for β gives $<\beta>B \in t$. This shows that $S_Y \subseteq \|<\beta>B\|$. Since $<\alpha><\beta>B \in \Gamma$ and $|s|\rho_\alpha Y$, (D2) for α then gives $<\alpha><\beta>B \in s$, so D-*Comp* yields $<\alpha;\beta>B \in s$ as desired.

Alternation.

(B1). Let A_s be a formula having

$$A_s \in t \quad \text{iff} \quad |s|\rho^+_{\alpha\cup\beta}|t|.$$

Using (B1) for α and β and the definition of $\rho^+_{\alpha\cup\beta}$, we get $A_s \in t$ whenever $s\overline{R_\alpha}t$ or $s\overline{R_\beta}t$. Hence $[\alpha]A_s, [\beta]A_s \in s$, so $[\alpha \cup \beta]A_s \in s$ by B-*Alt*.

(B2). If $|s|\overline{\rho_{\alpha\cup\beta}}|t|$, then either $|s|\overline{\rho_\alpha}|t|$ or else $|s|\overline{\rho_\beta}|t|$. Since B-*Alt* gives $[\alpha \cup \beta]B \in s$ only if $[\alpha]B, [\beta]B \in s$, (B2) for α and β then readily yield $\{B : [\alpha \cup \beta]B \in s \cap \Gamma\} \subseteq t$.

(D1). If $sR_{\alpha\cup\beta}T$, then by 10.22(3) there exists $W \subseteq T$ with $sR_\alpha W$ or $sR_\beta W$. Assuming (D1) for α and β, it follows that there is some $X \subseteq |W|$ with $|s|\rho_\alpha X$ or $|s|\rho_\beta X$. Hence $|s|\rho_{\alpha\cup\beta}X \subseteq |T|$.

(D2). Let $|s|\rho_{\alpha\cup\beta}X$, $S_X \subseteq \|B\|$, and $<\alpha\cup\beta>B \in \Gamma$. Then either $|s|\rho_\alpha X$ or $|s|\rho_\beta X$, and $<\alpha>B, <\beta>B \in \Gamma$. Hence by (D2) for α and β, one of $<\alpha>B$, and $<\beta>B$ is in s, implying $<\alpha\cup\beta>B \in s$ by D-*Alt*.

Combination.

(B1). Let A_s be a formula having

$$A_s \in t \quad \text{iff} \quad |s|\rho^+_{\alpha\cap\beta}|t|.$$

We show that

$$(<\alpha>\top \to [\beta]A_s),\ (<\beta>\top \to [\alpha]A_s) \in s, \tag{†}$$

which gives $[\alpha\cap\beta]A_s \in s$ by B-*Comb*.

To prove (†), let $<\alpha>\top \in s$. Then $sR_\alpha T$ for some T, and so by (D1) for α, $|s|\rho_\alpha X$ for some X. Then if $s\overline{R_\beta}t$ we have $|s|\rho^+_\beta|t|$ by (B1) for β, so with $|s|\rho_\alpha X$ we get $|s|\rho^+_{\alpha\cap\beta}|t|$, hence $A_s \in t$. This shows that $[\beta]A_s \in s$.

We have now shown that $(<\alpha>\top \to [\beta]A_s) \in s$. The proof that $(<\beta>\top \to [\alpha]A_s) \in s$ is similar.

(B2). Let $|s|\overline{\rho_{\alpha\cap\beta}}|t|$. Then there exist X, Y with $|s|\rho_\alpha X$, $|s|\rho_\beta Y$, and either $|t| \in X$ or $|t| \in Y$.

Now suppose $[\alpha\cap\beta]B \in s \cap \Gamma$. Then $<\alpha>\top, <\beta>\top \in \Gamma$. Since $S_X, S_Y \subseteq \|\top\|$, (D2) for α and β then give $<\alpha>\top, <\beta>\top \in s$. Hence axiom B-*Comb* implies $[\beta]B, [\alpha]B \in s$. But if $|t| \in X$, then $|s|\overline{\rho_\alpha}|t|$, so (B2) for α gives $B \in t$. If however $|t| \in Y$, we get the same conclusion from (B2) for β.

(D1). If $sR_{\alpha\cap\beta}T$, then by 10.22(4) there exist W_1, W_2 with $sR_\alpha W_1$, $sR_\beta W_2$, and $T = W_1 \cup W_2$. By (D1) for α and β, it follows that there exist X_1, X_2 with $|s|\rho_\alpha X_1 \subseteq |W_1|$ and $|s|\rho_\beta X_2 \subseteq |W_2|$. Hence

$$|s|\rho_{\alpha\cap\beta}(X_1 \cup X_2) \subseteq |W_1| \cup |W_2| \subseteq |T|.$$

(D2). Let $|s|\rho_{\alpha\cap\beta}X$, $S_X \subseteq \|B\|$, and $<\alpha\cap\beta>B \in \Gamma$. Then by definition of $\rho_{\alpha\cap\beta}$, there exist Y, Z with $|s|\rho_\alpha Y$, $|s|\rho_\beta Z$, and $X = Y \cup Z$. But $<\alpha>B, <\beta>B \in \Gamma$, and $S_Y, S_Z \subseteq S_X \subseteq \|B\|$, so by (D2) for α and β we get $<\alpha>B, <\beta>B \in s$. Axiom D-*Comb* then implies $<\alpha\cap\beta>B \in s$.

Iteration.

(B1). This is essentially as in the Ancestral Lemma 9.8. Let A_s be a formula having

$$A_s \in t \quad \text{iff} \quad |s|\rho^+_{\alpha^*}|t|.$$

We show that

$$\vdash A_s \to [\alpha]A_s. \tag{†}$$

For, if $t \in S^m$ and $A_s \in t$, then $|s|(\rho_\alpha^+)^*|t|$, and so $|s|(\rho_\alpha^+)^n|t|$ for some $n \geq 0$. Then if $t\overline{R_\alpha}u$, (B1) for α implies $|t|\rho_\alpha^+|u|$, hence $|s|(\rho_\alpha^+)^{n+1}|u|$, so $|s|\rho_{\alpha^*}^+|u|$, and therefore $A_s \in u$. This shows $[\alpha]A_s \in t$, as required for (†).

By the rule of Necessitation for $[\alpha^*]$ and axiom B-*Ind*, we then have $(A_s \to [\alpha^*]A_s) \in s$. But $A_s \in s$ as $|s|(\rho_\alpha^+)^0|s|$, so $[\alpha^*]A_s \in s$, yielding (B1) for α^*.

(B2). Since $\overline{\rho_{\alpha^*}} = \overline{\rho_\alpha^{(*)}} = (\overline{\rho_\alpha})^*$, we want to show that

$$|s|(\overline{\rho_\alpha})^*|t| \quad \text{implies} \quad \{B : [\alpha^*]B \in s \cap \Gamma\} \subseteq t.$$

Using (B2) for α and the $CPDL$-theorem $[\alpha^*]B \to [\alpha][\alpha^*]B$ (by B-*Mix*), we show, in similar fashion to 9.8 and 10.7, that for all $n \geq 0$,

$$|s|(\overline{\rho_\alpha})^n|t| \quad \text{implies} \quad \{[\alpha^*]B : [\alpha^*]B \in s \cap \Gamma\} \subseteq t.$$

Then if $|s|(\overline{\rho_\alpha})^*|t|$, we have $|s|(\overline{\rho_\alpha})^n|t|$ for some n, so if $[\alpha^*]B \in s \cap \Gamma$ then $[\alpha^*]B \in t$, hence $B \in t$ as $\vdash [\alpha^*]B \to B$ by B-*Mix*.

(D1). For any set $T \subseteq S^m$, let A_T be a formula such that for all $s \in S^m$,

$$A_T \in s \quad \text{iff} \quad |s|\rho_{\alpha^*}X \text{ for some } X \subseteq |T|.$$

We will prove

$$T \subseteq \|A_T\|, \tag{†}$$

and

$$\vdash <\alpha>A_T \to A_T. \tag{‡}$$

From these we derive (D1) for α^* as follows. If $sR_{\alpha^*}T$, then from (†) we get $<\alpha^*>A_T \in s$ (10.19(1)). But from (‡) by Necessitation for α^* and axiom D-*Ind*,

$$\vdash <\alpha^*>A_T \to A_T,$$

so $A_T \in s$, giving $|s|\rho_{\alpha^*}X$ for some $X \subseteq |T|$ as desired.

To prove (†), let $t \in T$. Then $|t|\rho_{\alpha^*}\{|t|\}$, since $Id \subseteq \rho_\alpha^{(*)} = \rho_{\alpha^*}$, and $\{|t|\} \subseteq |T|$, so with $X = \{|t|\}$ we fulfill $A_T \in t$, and hence $t \in \|A_T\|$.

For (‡) it suffices to show that any maximal set containing $<\alpha>A_T$ must also contain A_T. So, let $s \in S^m$ have $<\alpha>A_T \in s$. Then $sR_\alpha U$ for some $U \subseteq \|A_T\|$. By (D1) for α, $|s|\rho_\alpha X$ for some $X \subseteq |U|$. Thus for some $k \in \omega$ we have $X = \{|u_0|, \ldots, |u_{k-1}|\}$, for some $u_0, \ldots, u_{k-1} \in U$.

Now for each i with $0 \leq i < k$ we have $A_T \in u_i$, since $U \subseteq \|A_T\|$, and so $|u_i|\rho_{\alpha^*}Y_i$ for some $Y_i \subseteq |T|$. Since $\mathcal{M}_\Gamma$ is standard for α^*, it follows that

$|u_i|\rho_\alpha^{(n_i)}Y_i$ for some n_i. Let n be the maximum of $n_0, \ldots, n_{k-1}$. Then since the reachability relations $\rho_\alpha^{(m)}$ increase monotonically with m (Exercise 10.11(4)), we have $|u_i|\rho_\alpha^{(n)}Y_i$ for all $i < k$. Thus if $Y = \bigcup\{Y_i : 0 \leq i < k\}$, then $|s|(\rho_\alpha \cdot \rho_\alpha^{(n)})Y$, hence $|s|\rho_\alpha^{(n+1)}Y$, and so $|s|\rho_\alpha^{(*)}Y$. Therefore we have $|s|\rho_{\alpha^*}Y \subseteq |T|$, which ensures that $A_T \in s$ as desired.

(D2). If $|s|\rho_{\alpha^*}X$, then $|s|\rho_\alpha^{(n)}X$ for some n. Hence it suffices to prove that for all $n \geq 0$, and all $s \in S^m$,

$$\text{if } |s|\rho_\alpha^{(n)}X \text{ and } S_X \subseteq \|B\|, \text{ then } <\alpha^*>B \in \Gamma \text{ implies } <\alpha^*>B \in s. \qquad (\dagger)$$

For the case $n = 0$, if $|s|\rho_\alpha^{(0)}X$, i.e. $|s|Id\,X$, then $X = \{|s|\}$, so if $S_X \subseteq \|B\|$, then as $s \in S_X$ it follows that $B \in s$, and hence that $<\alpha^*>B \in s$ by axiom D-*Mix*.

Now make the inductive assumption that ($\dagger$) holds for n, and let $|s|\rho_\alpha^{(n+1)}X$, $S_X \subseteq \|B\|$, and $<\alpha^*>B \in \Gamma$. Then either $|s|\rho_\alpha^{(0)}X$, whence the desired result follows as above, or else $|s|(\rho_\alpha \cdot \rho_\alpha^{(n)})X$. In the latter case there must then be some Y with $|s|\rho_\alpha Y$ such that $X = \bigcup\{X_y : y \in Y\}$, with $y\rho_\alpha^{(n)}X_y$ for all $y \in Y$.

Then if $t \in S_Y$, we have $|t| \in Y$, so $S_{X_{|t|}} \subseteq S_X \subseteq \|B\|$, whence as $|t|\rho_\alpha^{(n)}X_{|t|}$, the hypothesis on n gives $<\alpha^*>B \in t$. Thus $S_Y \subseteq \|<\alpha^*>B\|$. But $<\alpha><\alpha^*>B \in \Gamma$, and $|s|\rho_\alpha Y$, so by (D2) for α, $<\alpha><\alpha^*>B \in s$. Hence by D-*Mix* we get our desideratum $<\alpha^*>B \in s$.

This show that ($\dagger$) holds for $n+1$, completing the inductive proof that it holds for all n, and hence completing the proof of Theorem 10.29.

Corollary 10.30. *$\mathcal{M}_\Gamma$ is a standard CPDL-model.*

Proof. By definition, $\mathcal{M}_\Gamma$ is standard except possibly for tests. Since it is a filtration of $\mathcal{M}^m$, the Filtration Lemma 10.27 then implies that

$$\rho_{B?} = \{(x, \{x\}) : \mathcal{M}_\Gamma \models_x B\}$$

for $B? \in Prog_\Gamma$, so that $\mathcal{M}_\Gamma$ is also standard for tests.

From this Corollary it follows in the usual way that any non-theorem of *CPDL* is falsifiable in a finite standard *CPDL*-model. Hence *CPDL* has the finite model property with respect to standard models, and is decidable.

Normality for $<\alpha>$

A natural condition to impose on models is that

$$sR_\alpha T \quad \text{implies} \quad T \neq \emptyset,$$

i.e.

$$\text{not-}sR_\alpha\emptyset,$$

since if $sR_\alpha T$ then T is the result of a terminating execution of α from s: termination implies the existence of a terminal state.

The corresponding axiom schema is

$$\text{D-}N: \qquad \neg\langle\alpha\rangle\bot,$$

which is always true under the binary relation semantics. Indeed it requires only the schema

$$[\alpha]\neg A \to \neg\langle\alpha\rangle A$$

to derive D-N from $[\alpha]\top$, and the latter is a theorem of any logic that is normal for $[\alpha]$.

Exercises 10.31

(1) Let Λ be a normal logic containing $CPDL$.

  (i) Show that relative to Λ, the schema D-N is equivalent to each of the schemata

$$[\alpha]\neg A \to \neg\langle\alpha\rangle A$$

$$\langle\alpha\rangle\neg A \to \neg[\alpha]A,$$

  i.e. Λ contains one of these three schemata if, and only if, it contains the others.

  (ii) Suppose that $\vdash_\Lambda \neg\langle\pi\rangle\bot$ for all atomic programs π. Prove that $\vdash_\Lambda \neg\langle\alpha\rangle\bot$ for all programs α.

  (iii) If $\vdash_\Lambda \neg\langle\alpha\rangle\bot$, then in the canonical model for Λ, not-$sR_\alpha\emptyset$.

(2) Show that in a standard model, if not-$sR_\pi\emptyset$ for all atomic π, then not-$sR_\alpha\emptyset$ for all α.

To prove the finite model property for the smallest normal logic obtained by adding D-N to $CPDL$, we modify the closure conditions on Γ to require that $\langle\pi\rangle\bot \in \Gamma$ whenever π occurs in Γ. Then in the finite filtration $\mathcal{M}_\Gamma$ it can be shown that not-$|s|\rho_\pi\emptyset$ for all atomic $\pi \in Prog_\Gamma$. To see this, observe that if $|s|\rho_\pi\emptyset$, then since $S_\emptyset = \emptyset = \|\bot\|$, property (D2) of ρ_π implies $\langle\pi\rangle\bot \in s$, which is inconsistent with D-N.

By Exercise 10.31(2) above, it then follows that not-$|s|\rho_\alpha\emptyset$ for all $\alpha \in Prog_\Gamma$, and so $\mathcal{M}_\Gamma$ is a D-N-model.

Sequential Atoms

The reachability relation R_α will be called *sequential* if

$$sR_\alpha T \quad \text{implies} \quad T = \{t\} \text{ for some } t.$$

The corresponding axiom schema is

$$Seq_\alpha : \quad [\alpha]\neg A \leftrightarrow \neg{<}\alpha{>}A,$$

from which $\neg{<}\alpha{>}\bot$ is derivable (10.31(1)(i)).

Lemma 10.32. *In the canonical model for a normal logic containing* $CPDL$ *and* Seq_α,

$$<\alpha>A \in s \quad \text{iff} \quad \textit{there exists } t \textit{ with } s\overline{R_\alpha}t \textit{ and } A \in t.$$

Proof. Recall that $s\overline{R_\alpha}t$ iff $s_\alpha \subseteq t$. Thus if $<\alpha>A \in s$, it suffices to show $s_\alpha \cup \{A\}$ is consistent. But if it were not, then $s_\alpha \vdash \neg A$, hence $[\alpha]\neg A \in s$ (10.18(5)), so $\neg{<}\alpha{>}A \in s$ by Seq_α, contrary to the consistency of s.

Conversely, if $s_\alpha \subseteq t$ and $A \in t$, then $\neg A \notin t$, so $[\alpha]\neg A \notin t$, whence by Seq_α and maximality of s, $<\alpha>A \in s$.

By a *sequential model* we will mean one in which the *atomic* relations R_π are sequential, so that parallelism depends on the presence of the combination connective $\alpha \cap \beta$ on programs. The (normal) logic determined by the class of sequential models is decidable, and is generated by adding the schemata Seq_π for all atomic π to $CPDL$. To show this, we modify the definition of ρ_π in $\mathcal{M}_\Gamma$, by defining it as the following sequential reachability relation on S_Γ.

$$x\rho_\pi\{y\} \quad \text{iff} \quad \exists s \in x\, \exists t \in y\, (s\overline{R_\pi}t).$$

Thus

$$x\rho_\pi^+ y \quad \text{iff} \quad x\overline{\rho_\pi}y \quad \text{iff} \quad \exists s \in x\, \exists t \in y\, (s_\pi \subseteq t),$$

from which it follows readily that ρ_π meets filtration conditions (B1) and (B2) (indeed the point is that ρ_π^+ is the smallest filtration of $\overline{R_\pi}$ in the sense of binary relation semantics).

To prove (D1) for ρ_π, let $sR_\pi T$ in the canonical model. Then $T \neq \emptyset$, since $\neg{<}\pi{>}\bot$ is derivable from Seq_π. Taking any $t \in T$, we get $s\overline{R_\pi}t$, and so $|s|\rho_\pi\{|t|\} \subseteq |T|$.

For (D2), let $|s|\rho_\pi X$, $S_X \subseteq \|B\|$, and $<\pi>B \in \Gamma$. Then there is some $s' \in |s|$ and some t such that $X = \{|t|\}$ and $s'\overline{R_\pi}t$. But then $t \in S_X$, so $B \in t$, and hence by Lemma 10.32, $<\pi>B \in s'$. Since $<\pi>B \in \Gamma$, we then get $<\pi>B \in s$ as desired.

This completes the proof that ρ_π is a Γ-filtration of R_π whenever $\pi \in Prog_\Gamma$. Thus $\mathcal{M}_\Gamma$ in this case is a finite sequential model that is a filtration of the canonical model. The rest of the story is as usual.

Further Studies

Dynamic logic is an extensive subject, with much to be learned by varying the class *Prog* of programs and its properties (cf. Harel [1984] and Kozen and Tiuryn [1989] for extensive surveys). One natural variation is to require atomic programs to be *deterministic*, so that, in terms of binary relation semantics, R_π becomes a *partial function* and the schema

$$<\pi>A \to [\pi]A$$

is valid (the quantificational logic of Part Three will have this property).

Now a logic Λ containing this schema will have a canonical model in which R_π^Λ is a partial function, but that feature will generally be lost in passing to a filtration $\mathcal{M}_\Gamma$. The problem of "unwinding" the atomic relations in $\mathcal{M}_\Gamma$ into functions, while preserving the standard-model conditions and the Filtration Lemma, is not easily solved. A solution is given in Ben-Ari, Halpern, and Pnueli [1982].

For an indication of the origin of dynamic logic, cf. Goldblatt [1986].

Part Three

First-Order Dynamic Logic

11 Assignments, Substitutions, and Quantifiers

In Part Three we study the language that results when the formalism of dynamic logic is added to a first-order language. The atomic programs π of PDL are replaced by *assignment commands* $(v := \sigma)$, where v is an individual variable, and σ a term. Such a command has the meaning "set v equal to σ", i.e. "assign to v the current value of σ", and is deterministic.

There is an intimate connection between the computational process of assignment to a variable, and the syntactic process of substitution for a variable. If A^v_σ is the result of replacing the free occurrences of v in a first-order formula A by σ, then

$$[v := \sigma]A \leftrightarrow A^v_\sigma$$

is valid. Because of this connection, we are able to use modal formulae of the form $[v := \sigma]A$ in places where the standard theory of first-order logic uses A^v_σ: it turns out that this is easier than trying to develop a theory of syntactic substitution in formulae that contain modal connectives.

In this context, the notion of *state* can be given a concrete interpretation. The current state of a computation is determined by saying what values all the variables currently have. Thus a state can be identified with a *valuation* of the individual variables, the same notion of valuation on which Tarski's definition of *satisfaction* in a first-order model is founded. Programs can then be interpreted as binary relations between valuations, and first-order dynamic logic becomes an enriched language for defining subsets of the space of valuations of a first-order model.

Defining an equivalence relation

$$s \sim_v t$$

to mean that states s and t differ only in the value they assign to v, we see that the Tarskian semantics translates to

$$\models_s \exists vA \quad \text{iff} \quad \text{for some state } t,\ s \sim_v t \text{ and } \models_t A;$$
$$\models_s \forall vA \quad \text{iff} \quad \text{for all states } t \text{ such that } s \sim_v t,\ \models_t A.$$

This makes $\exists v$ and $\forall v$ look like modal connectives, and indeed it is well known that formally they obey the laws of an $S5$-type $\Diamond$ and $\Box$. In fact we could (but won't) pursue this, and replace $\exists v$ and $\forall v$ altogether by $< v =? >$ and $[\, v =? \,]$, where the command $(v :=?)$ means "assign a *random* value to v" (Pratt [1976]).

Exercise 11.1

Explain informally why the following should be valid when v does not occur in σ.

$$< v := \sigma > A \leftrightarrow \exists v(v = \sigma \wedge A)$$
$$[\, v := \sigma \,] A \leftrightarrow \forall v(v = \sigma \rightarrow A)$$

Expressibility

The expressive power of first-order dynamic logic is greater than that of first-order logic itself. To see this, consider the following formula in the language of the arithmetic of natural numbers.

$$\forall w < v := 0; \textbf{while } v \neq w \textbf{ do } v := v + 1 > \top$$

This asserts that for all w, the displayed program has a terminating execution, i.e. that any w can be obtained by starting at 0 and repeatedly applying the successor operation $\zeta(n) = n + 1$. In other words: any set of numbers that contains 0 and is closed under ζ must contain everything. But this is a version of the Peano Induction Postulate, a postulate which cannot be expressed in the first-order language of the structure $(\omega, \zeta, 0)$. There is a single formula of dynamic logic which characterises this structure up to isomorphism, and from this it follows by standard arguments that the set of valid dynamic formulae is not effectively enumerable, unlike the first-order case (cf. Goldblatt [1982], §3.6, for details). This in turn means that there can be no adequate proof theory for first-order dynamic logic based on an enumerable set of axioms and an enumerable set of decidable inference rules. To develop a proof theory then, we will have to use *infinitary* rules of inference. The rule-schema we need is:

$$\textit{if} \vdash A \rightarrow [\,\beta; \alpha^n\,]B \textit{ for all } n \in \omega, \textit{ then} \vdash A \rightarrow [\,\beta; \alpha^*\,]B.$$

Exercises 11.2

(1) Verify that this rule preserves truth in standard PDL-models.

(2) The *Archimedean Property* of the real-number field $\mathbb{R}$ asserts that

$$\forall x\, \exists n \in \omega\, (x < n).$$

Express this as a sentence in the dynamic logic of an appropriate first-order structure based on $\mathbb{R}$.

(3) In the first-order dynamic logic of the language of groups, write a formula that expresses the notion of a *cyclic* group. Do the same for the notion of *divisible* group.

It would be possible to develop a theory in which $(v := \sigma)$ induces a *partial* function on states, allowing that evaluation of the term σ may fail to terminate. This would require the use of atomic formulae $(\sigma\downarrow)$, expressing "σ is defined", which would be true in precisely those states in which σ had a value. However for expository and paedogogical purposes, the system discussed in these notes is going to be kept as simple, and as close to standard first-order model theory, as possible. A version of the theory with partially defined terms is worked out in full in Goldblatt [1982], Chapter 3.

12 | Syntax and Semantics

Let $L = Rel_L \cup Fun_L \cup Con_L \cup Var_L$ be an *alphabet* made up of disjoint sets of relation symbols (with specified arities); function symbols (with specified arities); individual constants; and variables. The set Var_L of variables is assumed to be denumerable. The syntax of the first-order dynamic language generated by L is as follows.

Relation symbols:	$\boldsymbol{P} \in Rel_L$
Function symbols:	$\boldsymbol{f} \in Fun_L$
Constants:	$\boldsymbol{c} \in Con_L$
Variables:	$v \in Var_L$
Terms:	$\sigma \in Term_L$
Boolean formulae:	$\varphi \in Bool_L$
Formulae:	$A \in Fma_L$
Programs:	$\alpha \in Prog_L$

$$\varphi ::= \boldsymbol{P}(\sigma_1, \ldots, \sigma_n) \mid \sigma_1 = \sigma_2 \mid \perp \mid \varphi_1 \to \varphi_2$$

$$\sigma ::= v \mid \boldsymbol{c} \mid \boldsymbol{f}(\sigma_1, \ldots, \sigma_n)$$

$$A ::= \varphi \mid A_1 \to A_2 \mid [\alpha]A \mid \forall v A$$

$$\alpha ::= (v := \sigma) \mid \alpha_1 ; \alpha_2 \mid \alpha_1 \cup \alpha_2 \mid \alpha^* \mid \varphi?$$

(where $\boldsymbol{f}$ and $\boldsymbol{P}$ are n-ary).

Formulae of the type $\boldsymbol{P}(\sigma_1, \ldots, \sigma_n)$ and $\sigma_1 = \sigma_2$ are called *atomic*. Boolean formulae are truth-functional combinations of atomic formulae. *First-order* formulae are those that contain no modal connectives $[\alpha]$. Programs of the form $(v := \sigma)$ are *assignments* and all other types of program are called *structured*.

The restriction of test programs $\varphi?$ to Boolean formulae is realistic, since in practice a computer could not test the truth-value of a formula involving quantification of variables ranging over infinite sets, or subformulae of the form $[\alpha]A$ (which may assert that some program has a halting computation, for instance).

L-structures

Let $\mathfrak{A} = (X, I)$ be an *L-structure* in the usual sense, i.e. I is a function with domain L such that:

$$\text{for each } n\text{-ary } \boldsymbol{P} \in Rel_L,\ I(\boldsymbol{P}) \subseteq X^n;$$
$$\text{for each } n\text{-ary } \boldsymbol{f} \in Fun_L,\ I(\boldsymbol{f}) : X^n \to X;$$
$$\text{for each } \boldsymbol{c} \in Con_L,\ I(\boldsymbol{c}) \in X.$$

An $\mathfrak{A}$-*valuation* is a function $V : Var_L \to X$, assigning to each variable v a "value" $V(v)$ in X. Such a function extends in a unique way to $Term_L$, assigning a value $V(\sigma) \in X$ to each term σ. The set of all $\mathfrak{A}$-valuations will be denoted $S^{\mathfrak{A}}$.

If V and V' are $\mathfrak{A}$-valuations, we write

$$V \sim_v V'$$

to mean that V and V' are identical except (possibly) in the value they assign to v. The notation $V(v/x)$ denotes that $\mathfrak{A}$-valuation V' such that $V \sim_v V'$ and $V'(v) = x$.

Familiarity is assumed with the definition of the relation

$$\mathfrak{A} \models A[V]$$

of satisfaction of *first-order* formula A in $\mathfrak{A}$ by $\mathfrak{A}$-valuation V. In particular,

$$\mathfrak{A} \models \forall v A \quad \text{iff} \quad \text{for all } x \in X,\ \mathfrak{A} \models A[V(v/x)].$$

A standard procedure in first-order model-theory is to expand the alphabet L relative to a given L-structure $\mathfrak{A} = (X, I)$ by adding a new constant $\boldsymbol{c}_x$ for each $x \in X$. The resulting alphabet will be denoted $L_{\mathfrak{A}}$.The interpretation function I extends to $L_{\mathfrak{A}}$ by putting $I(\boldsymbol{c}_x) = x$. It will be convenient to continue to refer to the resulting $L_{\mathfrak{A}}$-structure as $\mathfrak{A}$.

Note that any $\mathfrak{A}$-valuation $V : Var_L \to X$ will assign a value $V(\sigma) \in X$ to any $L_{\mathfrak{A}}$-term, with, in particular, $V(\boldsymbol{c}_x) = x$.

Models

An *L-model* for dynamic logic is a structure

$$\mathcal{M} = (\mathfrak{A}, S, R, V)$$

where

- $\mathfrak{A}$ is an L-structure, as above;

- S is a non-empty set (of *states*);
- $V : S \to S^{\mathfrak{A}}$, i.e. V is a function assigning to each $s \in S$ an $\mathfrak{A}$-valuation $V_s : Var_L \to X$;
- R is a function assigning to each program $\alpha \in Prog_L$ a binary relation $R_\alpha \subseteq S \times S$.

For $s, t \in S$, we write

$$s(v/x)t$$

to mean that $V_t = V_s(v/x)$, i.e. that $V_t(v) = x$ and $V_t(w) = V_s(w)$ whenever $w \neq v$. More generally, we will use the notation

$$s(v/\sigma)t$$

to mean that $s(v/V_s(\sigma))t$, i.e. that V_s and V_t differ only in that $V_t(v) = V_s(\sigma)$.

The definition of the truth-relation

$$\mathcal{M} \models_s A$$

can now be given as follows.

$\mathcal{M} \models_s \varphi$	iff	$\mathfrak{A} \models \varphi[V_s]$
$\mathcal{M} \models_s A_1 \to A_2$	iff	$\mathcal{M} \models_s A_1$ implies $\mathcal{M} \models_s A_2$
$\mathcal{M} \models_s [\alpha]A$	iff	for all $t \in S$, $sR_\alpha t$ implies $\mathcal{M} \models_t A$
$\mathcal{M} \models_s \forall vA$	iff	for all $x \in X$, if $s(v/x)t$ then $\mathcal{M} \models_t A$

As usual, we write $\mathcal{M} \models A$ if $\mathcal{M} \models_s A$ for all $s \in S$.

Having Enough States

The model $\mathcal{M}$ will be said to *have enough states* if

for all $v \in Var_L$, $s \in S$, and $x \in X$, there exists $t \in S$ with $s(v/x)t$.

This condition is clearly going to be required if the quantifier $\forall v$ is to get its intended meaning "for all $x \in X$" at each state.

Exercises 12.1

(1) $\mathcal{M} \models \forall v(A \to B) \to (\forall vA \to \forall vB)$.

(2) If $\mathcal{M} \models A$, then $\mathcal{M} \models \forall vA$.

(3) If $\mathcal{M}$ has enough states, and A is *first-order*:
   (i) $\mathcal{M} \models_s A$ iff $\mathfrak{A} \models A[V_s]$;
   (ii) $\mathfrak{A} \models A$ implies $\mathcal{M} \models A$;
   (iii) If A is a *sentence* (no free variables), and $\mathcal{M} \models_s A$ for some $s \in S$, then $\mathcal{M} \models A$.

Standard Models

An L-model $\mathcal{M}$ is *standard* if it satisfies the following conditions:

- L includes a constant $\boldsymbol{c}_x$ for each $x \in X$, with $I(\boldsymbol{c}_x) = x$;
- $R_{v:=\sigma}$ is *serial*, i.e. for all $s \in S$ there exists $t \in S$ with $sR_{v:=\sigma}t$;
- if $sR_{v:=\sigma}t$, then $s(v/\sigma)t$, i.e. $V_t \sim_v V_s$ and $V_t(v) = V_s(\sigma)$;
- for structured programs, the standard-model conditions (as given in §10) all hold. In particular, for Boolean tests, this requires that

$$R_{\varphi?} = \{(s,s) : \mathfrak{A} \models \varphi[V_s]\}.$$

By considering assignments of the form $(v := \boldsymbol{c}_x)$, the first three of these conditions collectively imply that

a standard model has enough states,

and so standard models interpret $\forall$ correctly.

A standard model is, by definition, a model for a language of the form $L_{\mathfrak{A}}$ (or an extension of such a language), where $\mathfrak{A}$ is its first-order structure. We may also refer to a standard L-model as being *standard for* L.

Natural Models

The *natural model* of an L-structure $\mathfrak{A} = (X, I)$ is the $L_{\mathfrak{A}}$-model

$$\mathcal{M}^{\mathfrak{A}} = (\mathfrak{A}, S^{\mathfrak{A}}, R^{\mathfrak{A}}, V^{\mathfrak{A}}),$$

where

- $S^{\mathfrak{A}}$ is the set of all $\mathfrak{A}$-valuations $s : Var_L \to X$;
- for each $s \in S^{\mathfrak{A}}$, $V^{\mathfrak{A}}_s(v) = s(v)$;
- $sR^{\mathfrak{A}}_{v:=\sigma}t$ iff $t = s(v/s(\sigma))$;
- for structured programs, $R^{\mathfrak{A}}_\alpha$ is defined inductively by the standard-model condition on α.

Since $S^{\mathfrak{A}}$ includes all possible $\mathfrak{A}$-valuations, $\mathcal{M}^{\mathfrak{A}}$ is a standard model, and has enough states. Also, since $R^{\mathfrak{A}}_{v:=\sigma}$ is a functional relation,

$$\mathcal{M}^{\mathfrak{A}} \models <v := \sigma>A \leftrightarrow [v := \sigma]A.$$

In fact a standard model will always verify this schema, because of the standard-model conditions on assignments, even though it need not in general interpret $(v := \sigma)$ as a function. The reason for introducing the more abstract notion of standard model is that it is convenient to be able to realise states as entities (such as maximal sets of formulae) other than valuations. This gives greater freedom in constructing models.

We now examine the relationship between an L-model $\mathcal{M} = (\mathfrak{A}, S, R, V)$ and the associated natural model $\mathcal{M}^{\mathfrak{A}}$. Observe that for each state s in $\mathcal{M}$, the valuation V_s is a state in $\mathcal{M}^{\mathfrak{A}}$, with the value assigned by $V^{\mathfrak{A}}$ to variable v in the $\mathcal{M}^{\mathfrak{A}}$-state V_s being $V_s(v)$, which is the same as the value assigned to v in the $\mathcal{M}$-state s. That is:

$$V^{\mathfrak{A}}_{V_s} = V_s.$$

Exercises 12.2

(1) $s(v/x)t$ in $\mathcal{M}$ iff $V_s(v/x)V_t$ in $\mathcal{M}^{\mathfrak{A}}$.

(2) For any L-term σ,

$$V^{\mathfrak{A}}_{V_s}(\sigma) = V_s(\sigma),$$

so $s(v/\sigma)t$ in $\mathcal{M}$ iff $V_s(v/\sigma)V_t$ in $\mathcal{M}^{\mathfrak{A}}$.

(3) If A is a *Boolean* L-formula,

$$\mathcal{M} \models_s A \quad \text{iff} \quad \mathcal{M}^{\mathfrak{A}} \models_{V_s} A.$$

(4) If $\mathcal{M}$ has enough states, the result of Exercise 3 holds for all *first-order* A.

p-Morphism Lemma 12.3. *If $\mathcal{M}$ is a standard model, with underlying structure $\mathfrak{A}$, then the function $V : S \to S^{\mathfrak{A}}$ is a p-morphism from $\mathcal{M}$ to $\mathcal{M}^{\mathfrak{A}}$.*

Proof. What is meant by "p-morphism" here is that for each program α:

$$sR_\alpha t \quad \text{implies} \quad V_s R^{\mathfrak{A}}_\alpha V_t, \quad \text{and}$$
$$V_s R^{\mathfrak{A}}_\alpha u \quad \text{implies} \quad \exists t(sR_\alpha t \;\&\; V_t = u).$$

For an assignment $(v := \sigma)$, the standard-model conditions and the definition of $R^{\mathfrak{A}}_{v:=\sigma}$ yield

$$sR_{v:=\sigma}t \quad \text{implies} \quad V_s R^{\mathfrak{A}}_{v:=\sigma} V_t.$$

For the second condition, suppose $V_s R^{\mathfrak{A}}_{v:=\sigma} u$, with $u \in S^{\mathfrak{A}}$. Since $R_{v:=\sigma}$ is serial, there exists a $t \in S$ with $sR_{v:=\sigma}t$, so that

$$V_t = V_s(v/V_s(\sigma)) = u.$$

Thus the desired result holds when α is an assignment. The inductive cases for structured commands use the fact that both models are standard.

Exercise 12.4

Complete the proof of 12.3.

Theorem 12.5. *If $\mathcal{M}$ is standard, then for any L-formula A,*

(1) $\mathcal{M} \models_s A$ *iff* $\mathcal{M}^{\mathfrak{A}} \models_{V_s} A$.

(2) $V_s = V_t$ *implies* ($\mathcal{M} \models_s A$ *iff* $\mathcal{M} \models_t A$).

(3) *If* $\mathcal{M}^{\mathfrak{A}} \models A$ *then* $\mathcal{M} \models A$.

Proof. (2) and (3) are easy consequences of (1). (1) itself is proven by induction on the formation of A. The case of Boolean formulae is taken care of by Exercise 12.2(3), while the inductive case $A = [\alpha]B$ is taken care of by the p-Morphism Lemma 12.3 in the same manner as in propositional modal logic.

We treat only the case $A = \forall vB$ in detail, assuming the result for B. If $\forall vB$ is false at s in $\mathcal{M}$, then for some $x \in X$ and some t with $s(v/x)t$, $\mathcal{M} \not\models_t B$. Then $V_s(v/x)V_t$ (12.2(1)), and $\mathcal{M}^{\mathfrak{A}} \not\models_{V_t} B$ by hypothesis on B, so $\mathcal{M}^{\mathfrak{A}} \not\models_{V_s} \forall vB$.

Conversely, if $\forall vB$ is false at V_s in $\mathcal{M}^{\mathfrak{A}}$, then for some $x \in X$, and some $u \in S^{\mathfrak{A}}$ with $V_s(v/x)u$, $\mathcal{M}^{\mathfrak{A}} \not\models_u B$. But then $V_s R^{\mathfrak{A}}_{v:=\mathbf{c}_x} u$, so by the p-Morphism Lemma, $sR_{v:=\mathbf{c}_x}t$ for some t with $V_t = u$. Then $\mathcal{M} \not\models_t B$ and $s(v/x)t$, so $\mathcal{M} \not\models_s \forall vB$.

Corollary 12.6. *The classes of standard models and natural models determine the same logic.*

Quantifier/Assignment Lemma 12.7. *In a standard model $\mathcal{M}$,*

$$\mathcal{M} \models_s \forall vA \quad \text{iff} \quad \text{for all } x \in X,\ \mathcal{M} \models_s [v := \mathbf{c}_x]A.$$

Proof. If $\mathcal{M} \models_s \forall vA$, then $sR_{v:=\mathbf{c}_x}t$ implies $s(v/x)t$, so $\mathcal{M} \models_t A$ by the semantic clause for $\forall$. Hence $\mathcal{M} \models_s [v := \mathbf{c}_x]A$.

On the other hand, if $\mathcal{M} \not\models_s \forall vA$, then $\mathcal{M} \not\models_t A$ for some t such that $s(v/x)t$ for some $x \in X$. Then $V_t = V_s(v/x)$, so in the natural model $\mathcal{M}^{\mathfrak{A}}$, $V_s R^{\mathfrak{A}}_{v:=\mathbf{c}_x} V_t$. By the p-Morphism Lemma, there is an $\mathcal{M}$-state t' with $sR_{v:=\mathbf{c}_x}t'$ and $V_{t'} = V_t$. Then by Theorem 12.5(2), $\mathcal{M} \not\models_{t'} A$, so $\mathcal{M} \not\models_s [v := \mathbf{c}_x]A$.

Axioms

We now list some schemata, relating quantifiers to assignments, that will be used to axiomatise the logic of natural and standard models of a countable language. For this purpose, we denote by $VarA$ the (finite) set of all variables v that have an occurrence in A. Likewise, $Var\alpha$ is the set of variables occurring in program α.

A1: $\forall v(A \to B) \to (\forall vA \to \forall vB)$

A2: $A \to \forall vA$, for $v \notin VarA$

A3: $\forall vA \to [v := \sigma]A$

A4: $\forall w[v := w]A \rightarrow \forall vA,$ for $w \notin \{v\} \cup VarA$
A5: $\forall vA \rightarrow [v := \sigma]\forall vA$
A6: $\forall w[v := \sigma]A \rightarrow [v := \sigma]\forall wA,$ for $w \notin Var(v := \sigma)$
A7: $<v := \sigma>A \leftrightarrow [v := \sigma]A$
A8: $[v := \sigma]A \leftrightarrow A^v_\sigma$ for *atomic* A
A9: $[v := \sigma][v := \tau]A \rightarrow [v := \tau^v_\sigma]A$
A10: $[v := \sigma][w := \tau]A \rightarrow [w := \tau^v_\sigma][v := \sigma]A,$ for $w \notin Var(v := \sigma)$
A11: $\sigma = \tau \rightarrow ([v := \sigma]A \leftrightarrow [v := \tau]A)$

Note that A8 asserts the equivalence of assignment and substitution for atomic formulae only (cf. the role of this axiom and its consequences in the Assignment Lemma 14.3 in the last section).

Soundness

All of the schemata A1-A11 are true in standard models. A1 and A2 are familiar from first-order logic, although in the present computational context, A2 can be regarded as asserting that the truth-value of a formula A is not affected by an assignment to a variable that does not occur in A. A3 and A4 together give the equivalence of "for all v, A", and "after every assignment to v, A". A6 is an instance of the *Barcan formula*

$$\forall w \Box A \rightarrow \Box \forall w A$$

which figures prominently in studies of modal predicate logics. The Barcan formula is true when each possible world has the same domain of individuals associated with it, i.e. when the range of the quantifier $\forall w$ is independent of the world (state) in which truth is being evaluated. That property is satisfied by our present models.

The verification of A8 derives from the fact that for any $\mathfrak{A}$-valuation V, and any term τ,

$$V(\tau^v_\sigma) = V(v/V(\sigma))(\tau),$$

from which it can be shown that

$$\mathfrak{A} \models A^v_\sigma[V] \quad \text{iff} \quad \mathfrak{A} \models A[V(v/V(\sigma))],$$

for any first-order A (cf. e.g. the Substitution Lemma of Enderton [1972], p.127 for details). Hence

$$\mathcal{M}^{\mathfrak{A}} \models_s A^v_\sigma \quad \text{iff} \quad \mathcal{M}^{\mathfrak{A}} \models_{s(v/s(\sigma))} A \quad \text{iff} \quad \mathcal{M}^{\mathfrak{A}} \models_s [v := \sigma]A,$$

showing that A8 is true in natural models $\mathcal{M}^{\mathfrak{A}}$. But that is enough to make it true in all standard models, by Theorem 12.5(3).

Exercise 12.8

Suppose $s(v/\sigma)t$ in an $\mathfrak{A}$-based model $\mathcal{M}$.

(1) Show that $V_s(\tau^v_\sigma) = V_t(\tau)$, for any L-term τ.

(2) If $\mathcal{M}$ has enough states, show that for all first-order A,

$$\mathcal{M} \models_s A^v_\sigma \quad \text{iff} \quad \mathcal{M} \models_t A.$$

The intuitive meaning of the remaining axioms is left for the reader to ponder. Formal proofs of the truth of A1-A11 are tedious (although instructive) and will not be repeated here. Full details appear on pp.130-136 of Goldblatt [1982]. These proofs depend on some technical lemmas establishing that the truth of a formula A is not affected by an assignment to a variable not in $VarA$. These lemmas are given as

Exercises 12.9

(1) Suppose that $v \notin Var(\alpha)$. Then in $\mathcal{M}^{\mathfrak{A}}$, if $s(v/x)t$, then

$$tR^{\mathfrak{A}}_\alpha t' \quad \text{iff} \quad \exists s'(sR^{\mathfrak{A}}_\alpha s' \ \& \ s'(v/x)t')$$

(prove this by induction on α).

(2) Suppose that $A \in Fma_{L^{\mathfrak{A}}}$ and $v \notin VarA$. Then in $\mathcal{M}^{\mathfrak{A}}$, if $s(v/x)t$,

$$\mathcal{M}^{\mathfrak{A}} \models_s A \quad \text{iff} \quad \mathcal{M}^{\mathfrak{A}} \models_t A.$$

(3) If $v \notin VarA$, then in any standard model $\mathcal{M}$, if $sR_{v:=\mathbf{c}_x}t$, then

$$\mathcal{M} \models_s A \quad \text{iff} \quad \mathcal{M} \models_t A.$$

(4) Use these results to prove that A1-A11 are true in any standard model.

13 | Proof Theory

Axioms

The full set Axm_L of axioms for the first-order dynamic logic over an alphabet L comprises:

- all tautologies in Fma_L;
- the usual *Identity Axioms*

  $v = v$,

  $\sigma = \tau \to (A \to A')$, where A is *atomic*, and A' results by replacing some occurrences of σ by τ in A;

- the schemata *Comp*, *Alt*, *Mix*, and *Test* as for *PDL* in §10;
- the schemata A1-A11 from pages 151 and 152 in §12.

Rules

In addition to Detachment, and the Necessitation rule for each modal connective $[\alpha]$, the inference rule schemata we need are

Generalisation: from A deduce $\forall v A$;

Omega-Iteration: from $\{(A \to [\beta; \alpha^n]B) : n \in \omega\}$ deduce $(A \to [\beta; \alpha^*]B)$.

Note that we have left out the *PDL*-axiom

$$Ind : [\alpha^*](A \to [\alpha]A) \to (A \to [\alpha^*]A).$$

Its place has been taken by Omega-Iteration (cf. Exercise 13.1(10)) below.

Theorems

Let Λ_L be the smallest normal modal logic in Fma_L that contains Axm_L and is closed under Generalisation and Omega-Iteration, i.e.

$$A \in \Lambda_L \quad \text{implies} \quad \forall v A \in \Lambda_L;$$

$$\{(A \to [\beta; \alpha^n]B) : n \in \omega\} \subseteq \Lambda_L \quad \text{implies} \quad (A \to [\beta; \alpha^*]B) \in \Lambda_L.$$

The members of Λ_L are the *L-theorems.* If $A \in \Lambda_L$, we write $\vdash_L A$, or just $\vdash A$ if the context is understood. The main result of Part Three is that the theorems are precisely those formulae that are true in all natural models.

Exercises 13.1

The following are L-theorems.

(1) $\sigma = \sigma$.

(2) $\sigma = \tau \to \tau = \sigma$.

(3) $\sigma = \tau \to (\tau = \rho \to \sigma = \rho)$.

(4) $[v := \sigma]\neg A \leftrightarrow \neg[v := \sigma]A$.

(5) $[v := \sigma](A \to B) \leftrightarrow ([v := \sigma]A \to [v := \sigma]B)$.

(6) $[v := \sigma]\varphi \leftrightarrow \varphi^v_\sigma$, for any *Boolean* φ.

(7) $[\mathbf{skip}; \alpha]A \leftrightarrow [\alpha]A$.

(8) $[\alpha^n]A \leftrightarrow [\alpha]^n A$.

(9) $(A \to [\beta; \alpha^*]B) \to (A \to [\beta; \alpha^n]B)$.

(10) *Ind.* Hint: show that

$$\vdash B \to [\alpha]B,$$

where B is $A \wedge [\alpha^*](A \to [\alpha]A)$. Use Omega-Iteration to obtain

$$\vdash B \to [\alpha^*]B.$$

Theories and Deducibility

An *L-theory* is a set Δ of L-formulae that contains Λ_L and is closed under *Detachment* and *Omega-Iteration* (but not necessarily under Generalisation or Necessitation). If $\Gamma \cup \{A\} \subseteq Fma_L$, then A is *deducible from Γ in L*, $\Gamma \vdash_L A$, if A belongs to every L-theory that contains Γ. This type of definition appears as a theorem in the finitary proof theory of propositional modal logic (cf. Corollary 2.6), but since we are using an infinitary inference rule, the finitary definition of deducibility is no longer appropriate.

A set Γ is L-consistent if $\Gamma \nvdash_L \bot$.

Exercises 13.2

(1) If $A \in \Gamma$, then $\Gamma \vdash A$.

(2) If $\vdash A$, then $\Gamma \vdash A$.

(3) If $\Gamma \vdash A$ and $\Gamma \subseteq \Theta$, then $\Theta \vdash A$.

(4) If $\Gamma \vdash A$ and $\Gamma \vdash A \to B$, then $\Gamma \vdash B$.

(5) If $\Gamma \vdash A \to [\beta; \alpha^n]B$ for all $n \in \omega$, then $\Gamma \vdash A \to [\beta; \alpha^*]B$.

(6) Γ is consistent iff there is no A with $\Gamma \vdash A$ and $\Gamma \vdash \neg A$.

(7) If $\mathcal{M}$ is a standard model, then
$\vdash A$ implies $\mathcal{M} \models A$.

(8) If $\mathcal{M}$ is a standard model, then for any $\mathcal{M}$-state s,

$$\{A : \mathcal{M} \models_s A\}$$

is a consistent theory.

(9) If Γ is a theory, then:
  (i) $\top \in \Gamma$;
  (ii) (*Deductive Closure*) if $\Gamma \vdash A$, then $A \in \Gamma$;
  (iii) if $\Gamma \vdash A \to B$ and $A \in \Gamma$, then $B \in \Gamma$;
  (iv) Γ is consistent iff $\bot \notin \Gamma$ iff $\Gamma \neq Fma$;
  (v) $[\alpha^*]A \in \Gamma$ iff $\{[\alpha]^n A : n \in \omega\} \subseteq \Gamma$.

Lemma 13.3. *If* $\{[v := \sigma](A \to [\beta; \alpha^n]B) : n \in \omega\} \subseteq \Gamma$, *and* Γ *is a theory, then*

$$[v := \sigma](A \to [\beta; \alpha^*]B) \in \Gamma.$$

Proof. For all n, by the axiom K for $[v := \sigma]$, and use of axiom $Comp$, we get

$$([v := \sigma]A \to [(v := \sigma; \beta); \alpha^n]B) \in \Gamma.$$

By closure of Γ under Omega-Iteration, this gives

$$([v := \sigma]A \to [(v := \sigma; \beta); \alpha^*]B) \in \Gamma.$$

and hence by $Comp$,

$$([v := \sigma]A \to [v := \sigma][\beta; \alpha^*]B) \in \Gamma.$$

Exercise 13.1(5) then gives the desired result.

Deduction Theorem 13.4. $\Gamma \cup \{A\} \vdash B$ *iff* $\Gamma \vdash A \to B$.

Proof. (Note that for finitary proof theory this was an easy consequence of the definition of deducibility (Exercise 2.2(8)).

Suppose that $\Gamma \cup \{A\} \vdash B$. Let

$$\Delta = \{D : \Gamma \vdash A \to D\}.$$

We want $B \in \Delta$, so by our hypothesis it will suffice to show that Δ is a *theory* containing $\Gamma \cup \{A\}$.

Now since $D \to (A \to D)$ is a tautology, it is deducible from Γ, and this leads to $\Gamma \vdash A \to D$, hence $D \in \Delta$, in case that $D \in \Gamma$ or $\vdash D$. Similarly, using the tautology $A \to A$ we get $A \in \Delta$.

Next, to show that Δ is closed under Detachment, suppose D and $D \to E$ are in Δ. Then the tautology

$$(A \to D) \to ((A \to (D \to E)) \to (A \to E))$$

leads to $\Gamma \vdash A \to E$, as desired.

Finally, suppose

$$\{(D \to [\beta; \alpha^n]E) : n \in \omega\} \subseteq \Delta.$$

Then for all n,

$$\Gamma \vdash A \to (D \to [\beta; \alpha^n]E),$$

and so,

$$\Gamma \vdash A \wedge D \to [\beta; \alpha^n]E.$$

By Omega-Iteration (Exercise 13.2(5)), this gives

$$\Gamma \vdash A \wedge D \to [\beta; \alpha^*]E,$$

and ultimately that

$$(D \to [\beta; \alpha^*]E) \in \Delta.$$

This completes the proof that Δ is a theory, and hence the proof that $\Gamma \cup \{A\} \vdash B$ implies $\Gamma \vdash A \to B$. The converse is given as an exercise.

Corollary 13.5.

(1) $\{A_1, ..., A_n\} \vdash B$ *iff* $\vdash A_1 \wedge ... \wedge A_n \to B$.

(2) $\Gamma \cup \{A\}$ *is consistent iff* $\Gamma \nvdash \neg A$.

(3) $\Gamma \cup \{\neg A\}$ *is consistent iff* $\Gamma \nvdash A$.

Proof. Exercise.

Generalisation Lemma 13.6. *If the constant* $\mathbf{c}$ *does not occur in* A *or* B, *and*

$$\vdash A \to [v := \mathbf{c}]B,$$

then

$$\vdash A \to \forall v B.$$

Proof. In a finitary proof-theory, we would have a finite proof sequence ending in the first formula, and so we would first replace $\mathbf{c}$ throughout this sequence by some fresh variable. In the infinitary situation we could also have used proof sequences to define deducibility, but these would be infinite in length. Such an infinite sequence might use up all the variables, so some relettering might be necessary to "free one up" so that it could replace $\mathbf{c}$.

As it is, we are using a more abstract inductive definition of deducibility, but here we can still apply the relettering idea in a way that is, if anything, a little simpler to describe. So, pick a variable $w \notin \{v\} \cup VarA \cup VarB$, and let $y \mapsto y'$ be an injective mapping of $Var \cup \{\mathbf{c}\}$ into Var that has $\mathbf{c}' = w$, and $y' = y$ for y in the finite set $\{v\} \cup VarA \cup VarB$.

Since Var is infinite, such a function exists. For each formula D, let D' be the result of replacing each variable y in D by y'. Then the injective correspondence $D \mapsto D'$ maps axioms to axioms, and instances of rules (Detachment, Necessitation, Generalisation, Omega-Iteration) to instances of the same rules. Thus the set

$$\{D \in Fma_L : \vdash D'\}$$

must contain Λ_L, and so in particular contain the theorem

$$A \to [v := \mathbf{c}]B,$$

implying that

$$\vdash (A \to [v := \mathbf{c}]B)'.$$

By the hypothesis on $\mathbf{c}$ and the definition of the relettering $y \mapsto y'$, this means that

$$\vdash A \to [v := w]B.$$

Then by the Generalisation rule, axioms A1 and A2, and the fact that $w \notin Var A$, we get

$$\vdash A \to \forall w[v := w]B.$$

Axiom A4 then provides the desired conclusion.

Exercise 13.7

If $\vdash \forall v(A \to [\beta; \alpha^n]B)$, for all n, then $\vdash \forall v(A \to [\beta; \alpha^*]B)$.

Maximal Theories

An *L-maximal* theory is one that is L-consistent, and contains one of A and $\neg A$, for each L-formula A.

Exercises 13.8

If Γ is a maximal theory:

(1) $\bot \notin \Gamma$;

(2) exactly one of $A, \neg A$ belongs to Γ;

(3) $(A \to B) \in \Gamma$ iff $A \in \Gamma$ implies $B \in \Gamma$.

Rich Theories

If $\chi \subseteq Con_L$ is a set of L-constants, then an L-theory is *χ-rich* if it satisfies

if $\forall vB \notin \Gamma$, then for some $\mathbf{c} \in \chi$, $[v := \mathbf{c}]B \notin \Gamma$.

If this holds, χ may be called a set of "witnesses" for Γ in L.

Exercise 13.9

If Γ is a χ-rich theory, then

$$\forall v B \in \Gamma \quad \text{iff} \quad \text{for all } \mathbf{c} \in \chi,\ [v := \mathbf{c}]B \in \Gamma.$$

Witness Lemma 13.10. *If Γ is a χ-rich maximal L-theory, then for any L-term σ there is a witness $\mathbf{c} \in \chi$ with $(\sigma = \mathbf{c}) \in \Gamma$.*

Proof. Since $\vdash (\sigma = \sigma)$, $\neg(\sigma = \sigma) \notin \Gamma$, and so using Exercise 13.1(6), $[v := \sigma]\neg(\sigma = v) \notin \Gamma$. Axiom A3 then yields $\forall v \neg(\sigma = v) \notin \Gamma$, so by χ-richness, $\neg(\sigma = \mathbf{c}) \notin \Gamma$ for some $\mathbf{c} \in \chi$ with maximality then giving $(\sigma = \mathbf{c}) \in \Gamma$.

Adjoining Constants

In order to develop a completeness theorem, we follow the "Henkin method" used in first-order logic, and extend a given alphabet by adding new constants to serve as witnesses for rich theories. So, from now we fix an alphabet L, and let χ be a denumerable set disjoint from L. Form a new alphabet L^χ by adding the members of χ to the set of constants. First it needs to be checked that this process does not allow any new L-formulae to become deducible:

Exercise 13.11

Use a relettering technique, as in the proof of the Generalisation Lemma 13.6, to show that if $A \in Fma_L$, then

$$\vdash_L A \quad \text{iff} \quad \vdash_{L^\chi} A.$$

Theorem 13.12. *Let L be countable. Then if $\nvdash_L A$, there is an L^χ-theory that is χ-rich and L^χ-maximal, and does not contain A.*

Proof. Since L and χ are countable, there is an enumeration

$$A_0, A_1, \ldots, A_n, \ldots\ldots$$

of the set Fma_{L^χ} of all L^χ-formulae. Define an increasing sequence

$$\Delta_0 \subseteq \ldots \subseteq \Delta_n \subseteq \ldots\ldots$$

of *finite* sets as follows.

$$\Delta_0 = \{\neg A\}.$$

Assuming inductively that the finite set Δ_n has been defined, if

$$\Delta_n \vdash_{L^\chi} A_n,$$

we put

$$\Delta_{n+1} = \Delta_n \cup \{A_n\}.$$

Otherwise, when $\Delta_n \nvdash A_n$, we consider cases (working with deducibility in L^χ throughout).

Case 1: A_n is neither of the form $(B \to [\beta; \alpha^*]D)$, nor of the form $\forall v B$. Then put

$$\Delta_{n+1} = \Delta_n \cup \{\neg A_n\}.$$

Case 2: A_n is of the form $(B \to [\beta; \alpha^*]D)$. Then by Omega-Iteration (Exercise 13.2(5)), for some $m \in \omega$,

$$\Delta_n \nvdash B \to [\beta; \alpha^m]D.$$

Let

$$\Delta_{n+1} = \Delta_n \cup \{\neg(B \to [\beta; \alpha^m]D), \neg A_n\}$$

for some (say the least) such m.

Case 3: A_n has the form $\forall v B$. Then put

$$\Delta_{n+1} = \Delta_n \cup \{\neg[v := \boldsymbol{c}]B, \neg A_n\},$$

where $\boldsymbol{c}$ is some member of χ not appearing in A_n or in any member of Δ_n (since Δ_n is finite, and χ infinite, such $\boldsymbol{c}$ must exist).

This completes the definition of the Δ_n's. The desired L^χ-theory is

$$\Delta = \bigcup\{\Delta_n : n \in \omega\}.$$

It is evident that Δ contains all L^χ-theorems, for if $\vdash A_n$, then $\Delta_n \vdash A_n$, and $A_n \in \Delta_{n+1}$. Also, the construction ensures that if $A_n \notin \Delta$, then $\neg A_n \in \Delta$. To proceed further, it is necessary to show that each Δ_n is consistent. This is done by induction on n. For $n = 0$, Corollary 13.5(3) provides the result, since $\nvdash A$. Assuming that Δ_n is consistent, observe that if $\Delta \vdash A_n$, it must follow that $\Delta_n \nvdash \neg A_n$ (Ex. 13.2(6)), and so $\Delta_{n+1} = \Delta_n \cup \{A_n\}$ is consistent (Corollary 13.5(2)). If however $\Delta_n \nvdash A_n$, we have the three above cases to consider.

Case 1: Here, invocation of Corollary 13.5(3) again gives the consistency of $\Delta_{n+1} = \Delta_n \cup \{\neg A_n\}$.

Case 2: In this case, if Δ_{n+1} were not consistent, then by 13.5(3) and the Deduction Theorem 13.4 we would have

$$\Delta_n \vdash \neg(B \to [\beta; \alpha^m]D) \to A_n.$$

But

$$\vdash A_n \to (B \to [\beta; \alpha^m]D)$$

(cf. Exercise 13.1(9)), and so by tautological reasoning,

$$\Delta_n \vdash B \rightarrow [\beta; \alpha^m]D,$$

which contradicts the definition of m.

Case 3: If Δ_{n+1} is inconsistent in this case, then

$$\Delta_n \vdash \neg[v := \boldsymbol{c}]B \rightarrow A_n.$$

Since axiom A3 gives

$$\vdash A_n \rightarrow [v := \boldsymbol{c}]B,$$

this implies

$$\Delta_n \vdash [v := \boldsymbol{c}]B,$$

and hence

$$\vdash D \rightarrow [v := \boldsymbol{c}]B,$$

where D is the conjunction of the finitely many members of Δ_n. From our choice of $\boldsymbol{c}$, the Generalisation Lemma 13.6 then gives

$$\vdash D \rightarrow \forall v B,$$

and so $\Delta_n \vdash \forall v B$, contradicting the definition of this case.

Now that we know that each Δ_n is consistent, it follows readily that for each L^{χ}-formula B, exactly one of B and $\neg B$ is in Δ: otherwise, $B, \neg B \in \Delta_n$ for some n, contradicting Δ_n's consistency. From this follows the closure of Δ under Detachment, for if $B, (B \rightarrow D) \in \Delta$ but $D \notin \Delta$, the inconsistent set $\{B, (B \rightarrow D), \neg D\}$ would be contained in some Δ_n. Also, the construction makes Δ closed under Omega-Iteration, for if $(B \rightarrow [\beta; \alpha^*]D) \notin \Delta$, then for some m we get $\neg(B \rightarrow [\beta; \alpha^m]D) \in \Delta$, and hence $(B \rightarrow [\beta; \alpha^m]D) \notin \Delta$.

At this point Δ has been shown to be an L^{χ}-maximal theory, and hence to be deductively closed (Ex. 13.2(9)). Consistency of Δ now follows, since $\Delta \vdash \perp$ would imply $\perp \in \Delta$, and hence $\perp \in \Delta_n$ for some n. Finally, since $\neg A \in \Delta_0$, we have $A \notin \Delta$.

14 | Canonical Model and Completeness

Suppose L is a countable alphabet, and $\nvdash_L A$. Adjoining a denumerable set χ of constants to L, apply Theorem 13.12 to obtain a χ-rich maximal L^χ-theory s_A, with $A \notin s_A$. We use s_A to define a standard L^χ-model

$$\mathcal{M}^A = (\mathfrak{A}^A, S^A, R^A, V^A)$$

that falsifies A. The definition of $\mathcal{M}^A$ will take some time to develop.

The Diagram

We define the *diagram* of the structure $\mathfrak{A}^A$ to be the set $Diag_A$ of all atomic L^χ-sentences, and negations of atomic L^χ-sentences, that belong to s_A. Thus $Diag_A$ consists of all L^χ-formulae that belong to s_A of the form $\boldsymbol{P}(\sigma_1, \dots, \sigma_n)$ or $\sigma = \tau$, and the negations of such formulae, where the terms involved contain no variables (only constants and function letters). The members of $Diag_A$ will all be true in the L^χ-structure $\mathfrak{A}^A$, and give a complete specification of its algebraic relations.

The Structure

The definition of $\mathfrak{A}^A$ is the standard one used, as in the Henkin completeness proof for first-order logic, to build a first-order structure out of a maximal theory.

Define an equivalence relation on χ by putting

$$\boldsymbol{c} \sim \boldsymbol{d} \quad \text{iff} \quad (\boldsymbol{c} = \boldsymbol{d}) \in s_A$$

(by Exercises 13.1, this is indeed an equivalence). Let

$$\widetilde{\boldsymbol{c}} = \{\boldsymbol{d} : \boldsymbol{c} \sim \boldsymbol{d}\}$$

be the $\sim$-equivalence class of $\boldsymbol{c}$ and

$$X^A = \{\widetilde{\boldsymbol{c}} : \boldsymbol{c} \in \chi\}.$$

Then put $\mathfrak{A}^A = (X^A, I)$, where the interpretation function I is defined as follows:

- if $\boldsymbol{P}$ is an n-ary relation symbol,
  $I(\boldsymbol{P})(\widetilde{c_1}, \ldots, \widetilde{c_n})$ iff $\boldsymbol{P}(\boldsymbol{c}_1, \ldots, \boldsymbol{c}_n) \in s_A$;
- if $\boldsymbol{f}$ is an n-ary function symbol,
  $I(\boldsymbol{f})(\widetilde{c_1}, \ldots, \widetilde{c_n}) = \widetilde{\boldsymbol{c}}$ iff $(\boldsymbol{f}(\boldsymbol{c}_1, \ldots, \boldsymbol{c}_n) = \boldsymbol{c}) \in s_A$;
- if $\boldsymbol{d}$ is a constant,
  $I(\boldsymbol{d}) = \widetilde{\boldsymbol{c}}$ iff $(\boldsymbol{d} = \boldsymbol{c}) \in s_A$.

Note that in the last two cases, a suitable witness $\boldsymbol{c}$ always does exist, by the Witness Lemma 13.10. In the case that the constant $\boldsymbol{d}$ belongs to χ, we have more simply that $I(\boldsymbol{d}) = \widetilde{\boldsymbol{d}}$. Hence every member of $\mathfrak{A}^A$ is "named" by a constant from χ.

The State Set

S^A is defined to be the collection of all sets $s \subseteq Fma_{L^\chi}$ such that

- s is a χ-rich maximal L^χ-theory, and
- $Diag_A \subseteq s$.

Exercises 14.1

(1) $s_A \in S^A$.

(2) If B is an atomic L^χ-sentence, or the negation of such a sentence, then for any $s \in S^A$,

$$B \in s \quad \text{iff} \quad B \in s_A.$$

The Valuations

$$V_s^A(v) = \widetilde{\boldsymbol{c}} \quad \text{iff} \quad (v = \boldsymbol{c}) \in s.$$

Observe that for any $s \in S^A$ and any variable v, the Witness Lemma 13.10 guarantees that there is a $\boldsymbol{c} \in \chi$ with $(v = \boldsymbol{c}) \in s$.

Exercises 14.2

(1) For any L^χ-term σ, and $s \in S^A$,

$$V_s^A(\sigma) = \widetilde{\boldsymbol{c}} \quad \text{iff} \quad (\sigma = \boldsymbol{c}) \in s.$$

(2) For any Boolean L^χ-formula φ, and $s \in S^A$,

$$\mathfrak{A}^A \models \varphi[V_s^A] \quad \text{iff} \quad \varphi \in s.$$

In order to model assignments, we need a major preliminary result:

Assignment Lemma 14.3. *For any L^χ-assignment $(v := \sigma)$, if $s \in S^A$ then $s(v := \sigma) \in S^A$, where*

$$s(v := \sigma) = \{B \in Fma_{L^\chi} : [\,v := \sigma\,]B \in s\}.$$

Proof. First of all, $s(v := \sigma)$ contains all L^χ-theorems, for if $\vdash B$ then $\vdash [\,v := \sigma\,]B$ by Necessitation, so $[\,v := \sigma\,]B \in s$, hence $B \in s(v := \sigma)$.

Closure of $s(v := \sigma)$ under Detachment follows directly from the axiom

$$K: \quad [\,v := \sigma\,](B \rightarrow D) \rightarrow ([\,v := \sigma\,]B \rightarrow [\,v := \sigma\,]D),$$

and the closure of s under Detachment.

Closure of $s(v := \sigma)$ under Omega-Iteration is the substance of Lemma 13.3 on page 156.

Thus $s(v := \sigma)$ is an L^χ-theory, and so is deductively closed. Consistency now follows, for if $s(v := \sigma) \vdash \bot$ then $\bot \in s(v := \sigma)$, and so $[\,v := \sigma\,] \bot \in s$, contradicting the consistency of s, since $\vdash \neg[\,v := \sigma\,] \bot$ by axiom A7 (cf. Exercise 13.1(4)). A7 also implies

$$([\,v := \sigma\,]B \vee [\,v := \sigma\,]\neg B) \in s,$$

so for any formula B, one of B and $\neg B$ is in $s(v := \sigma)$. Hence $s(v := \sigma)$ is a *maximal* theory.

To prove $Diag_A \subseteq s(v := \sigma)$, observe that if $\varphi \in Diag_A$, then by axiom A8 (cf. Exercise 13.1(6)),

$$\vdash \varphi^v_\sigma \rightarrow [\,v := \sigma\,]\varphi.$$

But, by definition, φ contains no variables, so $\varphi^v_\sigma = \varphi \in s$. Hence $[\,v := \sigma\,]\varphi \in s$, giving $\varphi \in s(v := \sigma)$.

It remains to show that $s(v := \sigma)$ is χ-rich, i.e. that χ is a set of witnesses for $s(v := \sigma)$ in L^χ. This will use all the remainder of our axioms on quantifiers and assignments.

Suppose then that $\forall wB \notin s(v := \sigma)$, i.e. $[\,v := \sigma\,]\forall wB \notin s$. We want $[\,w := \mathbf{c}]B \notin s(v := \sigma)$, for some $\mathbf{c} \in \chi$. There are two cases.

Case 1: $w = v$. Then by A5, $\forall wB \notin s$. Since s is χ-rich, $[\,w := \mathbf{c}]B \notin s$ for some $\mathbf{c} \in \chi$. But the formula

$$[\,w := \sigma\,][\,w := \mathbf{c}]B \rightarrow [\,w := \mathbf{c}]B$$

is an instance of axiom A9, so gives

$$[\,w := \sigma\,][\,w := \mathbf{c}]B \notin s,$$

hence $[w := \boldsymbol{c}]B \notin s(w := \sigma) = s(v := \sigma)$, as desired.

Case 2: $w \neq v$. By the Witness Lemma 13.10, there is a $\boldsymbol{d} \in \chi$ with $(\sigma = \boldsymbol{d}) \in s$. Applying A11 gives

$$[v := \boldsymbol{d}]\forall w B \notin s.$$

Since w does not occur in $(v := \boldsymbol{d})$, the "Barcan formula" A6 then yields

$$\forall w[v := \boldsymbol{d}]B \notin s,$$

so for some witness $\boldsymbol{c} \in \chi$,

$$[w := \boldsymbol{c}][v := \boldsymbol{d}]B \notin s.$$

But as an instance of A10 we have

$$[v := \boldsymbol{d}][w := \boldsymbol{c}]B \to [w := \boldsymbol{c}][v := \boldsymbol{d}]B,$$

so that we can conclude

$$[v := \boldsymbol{d}][w := \boldsymbol{c}]B \notin s.$$

Axiom A11 again then yields

$$[v := \sigma][w := \boldsymbol{c}]B \notin s,$$

whence $[w := \boldsymbol{c}]B \notin s(v := \sigma)$.

Corollary 14.4.

$$[v := \sigma]B \in s \quad \textit{iff} \quad B \in s(v := \sigma).$$

Proof. If $[v := \sigma]B \notin s$, then by A7, $[v := \sigma]\neg B \in s$, so $\neg B \in s(v := \sigma)$.

Modelling Programs

For assignments, we put

$$sR^A_{v:=\sigma}t \quad \text{iff} \quad t = s(v := \sigma),$$

while for structured commands, R^A_α is defined inductively by the standard-model condition on α. In particular, for Boolean tests,

$$R^A_{\varphi?} = \{(s,s) : \mathfrak{A}^A \models \varphi[V^A_s]\}.$$

This completes the definition of the $L\chi$-model $\mathcal{M}^A$.

Lemma 14.5. *$\mathcal{M}$ is a standard model.*

Proof. The only standard-model conditions not built in to the definition are the ones for assignments. For these, note first that the Assignment Lemma 14.3 ensures that $R^A_{v:=\sigma}$ is serial, i.e. that for all $s \in S^A$ there exists $t \in S^A$ with $sR^A_{v:=\sigma}t$.

Next, suppose that $sR^A_{v:=\sigma}t$. We have to show that $s(v/\sigma)t$, i.e. that V^A_s and V^A_t differ only in that $V^A_t(v) = V^A_s(\sigma)$. Let $V^A_s(\sigma) = \widetilde{\mathbf{c}}$, so that $(\sigma = \mathbf{c}) \in s$. Now by A8,

$$\vdash (\sigma = \mathbf{c}) \to [\, v := \sigma\,](v = \mathbf{c}),$$

so $[\, v := \sigma\,](v = \mathbf{c}) \in s$, whence

$$(v = \mathbf{c}) \in s(v := \sigma) = t,$$

and so $V^A_t(v) = \widetilde{\mathbf{c}} = V^A_s(\sigma)$, as required. But if w is any variable other than v, A8 gives

$$\vdash (w = \mathbf{c}) \to [\, v := \sigma\,](w = \mathbf{c}),$$

from which similar reasoning shows that if $V^A_s(w) = \widetilde{\mathbf{c}}$, then $V^A_t(w) = \widetilde{\mathbf{c}}$. This completes the proof.

Lemma 14.5 ensures that $\mathcal{M}^A$ has enough states, and so interprets the quantifier $\forall$ correctly. Moreover, from the Quantifier/Assignment Lemma 12.7, it gives

Corollary 14.6.

$$\mathcal{M}^A \models_s \forall v B \quad \text{iff} \quad \text{for all } \mathbf{c} \in \chi,\ \mathcal{M}^A \models_s [\, v := \mathbf{c}]B.$$

We are heading towards a Truth Lemma for $\mathcal{M}^A$, and, as a final preliminary, we extract a part of its proof for separate consideration. To this end, a formula B is defined to be *correct* if for every $s \in S^A$,

$$\mathcal{M}^A \models_s B \quad \text{iff} \quad B \in s.$$

Program Lemma 14.7. *Let α be an L^χ-program. Then for any L^χ-formula B, if B is correct, then $[\,\alpha\,]B$ is correct.*

Proof. By induction on the formation of α. Take first the case that α is an assignment $(v := \sigma)$. If B is correct, then in particular

$$\mathcal{M}^A \models_{s(v:=\sigma)} B \quad \text{iff} \quad B \in s(v := \sigma).$$

But by definition of $\mathcal{M}^A$,

$$\mathcal{M}^A \models_s [v := \sigma]B \quad \text{iff} \quad \mathcal{M}^A \models_{s(v:=\sigma)} B,$$

while by Corollary 14.4,

$$[v := \sigma]B \in s \quad \text{iff} \quad B \in s(v := \sigma),$$

and so $[v := \sigma]B$ is correct.

Next the case of a test $\varphi?$. The Boolean formula φ is correct (Exercise 14.2(2)), so that if B is correct it follows readily that $(\varphi \to B)$ is correct also. Correctness of $[\varphi?]B$ is then obtained by use of the formula

$$[\varphi?]B \leftrightarrow (\varphi \to B),$$

which is true in the standard model $\mathcal{M}^A$, and a member of every $s \in S^A$, since it is an instance of the axiom schema *Test*.

Now for the case of a program $\alpha;\beta$, under the inductive assumption that the Lemma holds for α and for β. Then if B is correct, the hypothesis on β makes $[\beta]B$ correct, and so the hypothesis on α applied to $[\beta]B$, makes $[\alpha][\beta]B$ correct, i.e.

$$\mathcal{M}^A \models_s [\alpha][\beta]B \quad \text{iff} \quad [\alpha][\beta]B \in s.$$

Correctness of $[\alpha;\beta]B$ then follows by using the instance

$$[\alpha;\beta]B \leftrightarrow [\alpha][\beta]B$$

of axiom *Comp*, which is true in the standard model $\mathcal{M}^A$.

The case of a program of the form $\alpha \cup \beta$ is similar to that of $\alpha;\beta$, using the axiom *Alt*, and is left to the reader.

Finally the case of an iterative program α^*, assuming the result for α. Suppose B is correct. First we show that $[\alpha]^n B$ is correct for all $n \in \omega$. If $n = 0$, this is just the assumption on B. Assuming that $[\alpha]^n B$ is correct, the hypothesis on α then gives $[\alpha][\alpha]^n B$, i.e. $[\alpha]^{n+1}B$, correct. Hence, by induction on n, we get

$$\mathcal{M}^A \models_s [\alpha]^n B \quad \text{iff} \quad [\alpha]^n B \in s,$$

for all n and s. But in the standard model $\mathcal{M}^A$,

$$\mathcal{M}^A \models_s [\alpha^*]B \quad \text{iff} \quad \text{for all } n \in \omega,\ \mathcal{M}^A \models_s [\alpha]^n B$$

(Exercise 10.1(1)), while

$$[\alpha^*]B \in s \quad \text{iff} \quad \text{for all } n \in \omega,\ [\alpha]^n B \in s,$$

by closure of s under Omega-Iteration etc. (Exercise 13.2(9)(v)). Hence $[\alpha^*]B$ is correct.

Truth-Lemma for $\mathcal{M}^A$. *Every L^χ-formula B is correct, i.e. for every $s \in S^A$,*

$$\mathcal{M}^A \models_s B \quad \text{iff} \quad B \in s.$$

Proof. That Boolean formulae are correct is Exercise 14.2(2). The truth-functional cases are as usual.

If B is correct, then for any program α, correctness of $[\alpha]B$ is given by the Program Lemma 14.7 (which was treated separately because it requires an "inner" induction on α).

Finally, consider $\forall vB$, assuming B is correct. We have

$$\mathcal{M}^A \models_s \forall vB \quad \text{iff} \quad \text{for all } \boldsymbol{c} \in \chi,\ \mathcal{M}^A \models_s [v := \boldsymbol{c}]B,$$

by Corollary 14.6, while for each $\boldsymbol{c} \in \chi$, the Program Lemma gives

$$\mathcal{M}^A \models_s [v := \boldsymbol{c}]B \quad \text{iff} \quad [v := \boldsymbol{c}]B \in s.$$

Since χ-richness and axiom A3 yield

$$\forall vB \in s \quad \text{iff} \quad \text{for all } \boldsymbol{c} \in \chi,\ [v := \boldsymbol{c}]B \in s,$$

correctness of $\forall vB$ then follows.

Completeness Theorem. *If L is countable, then for any L-formula A, the following are equivalent.*

(1) $\vdash_L A$.

(2) A is true in all natural L-models.

(3) A is true in all standard L-models.

Proof.
(1) *implies* (2): if $\mathcal{M}$ is natural, $\{A : \mathcal{M} \models A\}$ is a normal modal logic containing all axioms and closed under Generalisation and Omega-Iteration, hence containing

$$\Lambda_L = \{A : \vdash A\}.$$

(2) *implies* (3): Corollary 12.6 (from Theorem 12.5(3)).
(3) *implies* (1): if $\nvdash A$, then in the standard model $\mathcal{M}^A$ constructed above, the Truth Lemma gives A false at s_A.

Bibliography

In addition to the books and papers cited in the text, the following list includes other items of potential interest to the student of modal and temporal logic.

Ben-Ari, M., Halpern, J.Y., and Pnueli, A.
[1982] Deterministic propositional dynamic logic: finite models, complexity, and completeness, *J. Comp. Syst. Sci.*, **25**, 402–417.

Ben-Ari, M., Pnueli, A., and Manna Z.
[1983] The temporal logic of branching time, *Acta Informatica*, **20**, 207–226.

Blok, W.J.
[1980] The lattice of modal algebras: an algebraic investigation, *J. Symbolic Logic*, **45**, 221–236.

Boolos, George
[1979] *The Unprovability of Consistency*, Cambridge University Press.

Boolos, George, and Sambin, Giovanni
[1985] An incomplete system of modal logic, *J. Philosophical Logic 14*, 351–358.

Bull, R.A.
[1966] That all normal extensions of S4.3 have the finite model property, *Zeit. Math. Logik Grund. Math.*, **12**, 341–344.

Bull, Robert A., and Segerberg, Krister
[1984] Basic modal logic, in Gabbay and Guenthner (eds.), 1–88.

Burgess, John P.
[1984] Basic tense logic, in Gabbay and Guenthner (eds.), 89–133.

Chellas, Brian F.
[1980] *Modal Logic: An Introduction*, Cambridge University Press.

Clarke, E.M., and Emerson, E.A.
[1981] Design and synthesis of synchronisation skeletons using branching time temporal logic, in *Logics of Programs*, D. Kozen (ed.), Lecture Notes in Computer Science **131**, Springer-Verlag, 52–71.

[1982] Using branching time temporal logic to synthesize synchronisation skeletons, *Science of Computer Programming*, **2**, 241–266.

Cresswell, M.J.

[1984] An incomplete decidable modal logic, *J. Symbolic Logic*, **49**, 520–527.

de Bakker, J.W., de Roever, W.-P., and Rozenberg, G. (eds.)

[1989] *Linear Time, Branching Time, and Partial Order in Logics and Models for Concurrency*, Lecture Notes in Computer Science **354**, Springer-Verlag.

Emerson, E.A., and Halpern, Joseph Y.

[1985] Decision procedures and expressiveness in the temporal logic of branching time, *J. Computer and Systems Sciences*, **30**, 1–24.

Fine, K.

[1971] The logics containing S4.3, *Zeit. Math. Logik Grund. Math.*, **17**, 371–376.

[1974] An incomplete logic containing S4, *Theoria*, **40**, 23–29.

[1975] Some connections between modal and elementary logic, in *Proc. Third Scandinavion Logic Symposium*, Stig Kanger (ed.), Studies in Logic **82**, North-Holland, 15–31.

[1975i] Normal forms in modal logic, *Notre Dame J. of Formal Logic*, **16**, 229–237.

Fischer, M.J., and Ladner, R.F.

[1979] Propositional dynamic logic of regular programs, *J. Comp. Syst. Sci.*, **18**, 194–211.

Gabbay, D., and Guenthner, F. (eds.)

[1984] *Handbook of Philosophical Logic, Volume II: Extensions of Classical Logic*, D. Reidel.

Gabbay, D., Pnueli, A., Shelah, S., and Stavi, J.

[1980] On the temporal analysis of fairness, *Proc. 7th ACM Symp. on Principles of Programming Languages*, Las Vegas, Jan. 1980, 163–173.

Galton, Antony

[1987] *Temporal Logics and their Applications*, Academic Press.

Gödel, Kurt

[1933] Eine Interpretation des intuitionistischen Aussagenkalküls, *Ergebnisse eines mathematischens Kolloquiums*, 4 (1931-32), 39-40. English translation in *Kurt Gödel, Collected Works, vol. I*, Solomon Feferman et. al. (eds.), Oxford University Press, 1986, 296-303.

Goldblatt, Robert

[1975] First-order definability in modal logic, *J. Symbolic Logic*, **40**, 35–40.

[1975i] Solution to a completeness problem of Lemmon and Scott, *Notre Dame J. of Formal Logic*, **16**, 405–408.

[1976] Metamathematics of modal logic, *Reports on Mathematical Logic*, Polish Scientific Publishers, Warsaw-Cracow, **6**, 41–78 (Part I); and **7**, 21–52 (Part II).

[1980] Diodorean modality in Minkowski spacetime, *Studia Logica*, **39**, 219–236.

[1982] *Axiomatising the Logic of Computer Programming*, Lecture Notes in Computer Science **130**, Springer-Verlag.

[1982i] The semantics of Hoare's iteration rule, *Studia Logica*, **41**, 141–158.

[1986] Review of Fischer and Ladner [1979], Pratt [1976], Segerberg [1982], and other papers, *J. Symbolic Logic*, **51**, 225–227.

[1990] On closure under canonical embedding algebras, in *Algebraic Logic*, H. Andreka, J.D. Monk, and I. Nemeti (eds.), Colloquia Mathematica Societatis Janos Bolyai, **54**, North-Holland Publishing Co., 217–229.

[1991] The McKinsey axiom is not canonical, *J. Symbolic Logic*, **56**, 554–562.

Hailpern, Brent T.

[1982] *Verifying Concurrent Processes Using Temporal Logic*, Lecture Notes in Computer Science **129**, Springer-Verlag.

Harel, David

[1979] *First-order Dynamic Logic*, Lecture Notes in Computer Science **68**, Springer-Verlag.

[1984] Dynamic logic, in Gabbay and Guenthner (eds.), 497–604.

Hughes, G.E., and Cresswell, M.J.

[1968] *An Introduction to Modal Logic*, Methuen.

[1984] *A Companion to Modal Logic*, Methuen.

Kamp, J.A.W.

[1968] Tense Logic and the Theory of Order, Ph.D dissertation, UCLA.

Kozen, Dexter, and Tiuryn, Jerzy

[1989] Logics of Programs, Technical Report 89–962, Department of Computer Science, Cornell University. Published in *Handbook of Theoretical Computer Science, Vol. B*, North-Holland, 1990.

Kripke, Saul A.
[1959] A completeness theorem in modal logic, *J. Symbolic Logic*, **24**, 1–14.
[1963] Semantic analysis of modal logic I: normal propositional calculi, *Zeit. Math. Logik Grund. Math.*, **9**, 67–96.

Lemmon, E.J.
[1977] *An Introduction to Modal Logic*, in collaboration with Dana Scott, American Philosophical Quarterly Monograph Series **11**, Basil Blackwell, Oxford.

Lewis, C.I., and Langford, C.H.
[1932] *Symbolic Logic*, The Century Co.

Makinson, D.
[1969] A normal modal calculus between T and S4 without the finite model property, *J. Symbolic Logic*, **34**, 35–38.

Manna, Z., and Pnueli, A.
[1981] Verification of concurrent programs: the temporal framework, in *The Correctness Problem in Computer Science*, R.S. Boyer and J.S. Moore (eds), Academic Press, 215–273.

Moszkowski, Ben
[1986] *Executing Temporal Logic Programs*, Cambridge University Press.

Nerode, A., and Wijesekera, D.
[1990] Constructive concurrent dynamic logic I, Technical Report '90-43, Mathematical Sciences Institute, Cornell University.

Parikh, R.
[1984] Logics of knowledge, games, and dynamic logic, in *Foundations of Software Technology and Theoretical Computer Science*, Lecture Notes in Computer Science **181**, M.Joseph and R.Shyamasundar (eds.), 202–222.

Peleg, David
[1987] Concurrent dynamic logic, *JACM*, **34**, 450–479.
[1987i] Communication in concurrent dynamic logic, *J. Comp. Syst. Sci.*, **35**, 23–58.

Pnueli, A.
[1981] The temporal semantics of concurrent programs, *Theoretical Computer Science*, **13**, 45–60.

Pratt, V.R.
[1976] Semantical considerations on Floyd-Hoare logic, *Proc. 17th IEEE Symp. on Foundations of Computer Science*, 109–121.

Prior, Arthur
[1957] *Time and Modality*, Clarendon Press, Oxford.
[1967] *Past, Present, and Future*, Clarendon Press, Oxford.

Rescher N., and Urquhart, A.
[1971] *Temporal Logic*, Springer-Verlag.

Rosenchein, Stanley J.
[1985] Formal theories of knowledge in AI and robotics, *New Generation Computing*, **3**, Oshma Ltd., Tokyo. Also as Technical Note 362, Artificial Intelligence Center, SRI International, Menlo Park, California.

Rosenchein, Stanley J., and Kaelbling, Leslie Pack
[1986] The synthesis of digital machines with provable epistemic properties, SRI International and CSLI Stanford.

Sahlqvist, H.,
[1975] Completeness and correspondence in first and second order semantics for modal logic, in *Proceedings of the Third Scandinavian Logic Symposium*, ed. Stig Kanger, North-Holland, 110–143.

Sambin, G., and Vaccaro, V.,
[1989] A new proof of Sahlqvist's theorem on modal definability and completeness, *J. Symbolic Logic*, **54**, 992–999.

Segerberg, Krister
[1970] Modal logics with linear alternative relations, *Theoria*, **36**, 301–322.
[1971] *An Essay in Classical Modal Logic*, Philosophical studies published by the Philosophical Society and the Department of Philosophy, University of Uppsala, volume **13**, Uppsala.
[1982] A completeness theorem in the modal logic of programs, in *Universal Algebra and Applications*, T. Traczyk (ed.), Banach Centre Publications **9**, PWN - Polish Scientific Publishers, Warsaw, 31–46.

Thomason, S.K.
[1972] Semantic analysis of tense logics, *J. Symbolic Logic*, **37**, 150–158.
[1974] An incompleteness theorem in modal logic, *Theoria*, **40**, 30–34.
[1975] Reduction of second-order logic to modal logic, *Zeit. Math. Logik Grund. Math.*, **21**, 107–114.

Urquhart, A.
[1981] Decidability and the finite model property, *J. Philosophical Logic*, **10**, 367–370.

van Benthem, J.F.A.K.,

[1975] A note on modal formulas and relational properties, *J. Symbolic Logic*, **40**, 55–58.

[1978] Two simple incomplete modal logics, *Theoria*, **44**, 25–37.

[1980] Some kinds of modal completeness, *Studia Logica*, **39**, 125–141.

[1983] *The Logic of Time*, D. Reidel.

Index

CSLI Publications

Reports

The following titles have been published in the CSLI Reports series. These reports may be obtained from CSLI Publications, Ventura Hall, Stanford University, Stanford, CA 94305-4115.

Coordination and How to Distinguish Categories Ivan Sag, Gerald Gazdar, Thomas Wasow, and Steven Weisler CSLI–84–3 ($*3.50*)

Belief and Incompleteness Kurt Konolige CSLI–84–4 ($*4.50*)

Equality, Types, Modules and Generics for Logic Programming Joseph Goguen and José Meseguer CSLI–84–5 ($*2.50*)

Lessons from Bolzano Johan van Benthem CSLI–84–6 ($*1.50*)

Self-propagating Search: A Unified Theory of Memory Pentti Kanerva CSLI–84–7 ($*9.00*)

Reflection and Semantics in LISP Brian Cantwell Smith CSLI–84–8 ($*2.50*)

The Implementation of Procedurally Reflective Languages Jim des Rivières and Brian Cantwell Smith CSLI–84–9 ($*3.00*)

Parameterized Programming Joseph Goguen CSLI–84–10 ($*3.50*)

Shifting Situations and Shaken Attitudes Jon Barwise and John Perry CSLI–84–13 ($*4.50*)

Completeness of Many-Sorted Equational Logic Joseph Goguen and José Meseguer CSLI–84–15 ($*2.50*)

Moving the Semantic Fulcrum Terry Winograd CSLI–84–17 ($*1.50*)

On the Mathematical Properties of Linguistic Theories C. Raymond Perrault CSLI–84–18 ($*3.00*)

A Simple and Efficient Implementation of Higher-order Functions in LISP Michael P. Georgeff and Stephen F.Bodnar CSLI–84–19 ($*4.50*)

On the Axiomatization of "if-then-else" Irène Guessarian and José Meseguer CSLI–85–20 ($*3.00*)

The Situation in Logic–II: Conditionals and Conditional Information Jon Barwise CSLI–84–21 ($*3.00*)

Principles of OBJ2 Kokichi Futatsugi, Joseph A. Goguen, Jean-Pierre Jouannaud, and José Meseguer CSLI–85–22 ($*2.00*)

Querying Logical Databases Moshe Vardi CSLI–85–23 ($*1.50*)

Computationally Relevant Properties of Natural Languages and Their Grammar Gerald Gazdar and Geoff Pullum CSLI–85–24 ($*3.50*)

An Internal Semantics for Modal Logic: Preliminary Report Ronald Fagin and Moshe Vardi CSLI–85–25 ($*2.00*)

The Situation in Logic–III: Situations, Sets and the Axiom of Foundation Jon Barwise CSLI–85–26 ($*2.50*)

Semantic Automata Johan van Benthem CSLI–85–27 ($*2.50*)

Restrictive and Non-Restrictive Modification Peter Sells CSLI–85–28 ($*3.00*)

Institutions: Abstract Model Theory for Computer Science J. A. Goguen and R. M. Burstall CSLI–85–30 ($*4.50*)

A Formal Theory of Knowledge and Action Robert C. Moore CSLI–85–31 ($*5.50*)

Finite State Morphology: A Review of Koskenniemi (1983) Gerald Gazdar CSLI–85–32 ($*1.50*)

The Role of Logic in Artificial Intelligence Robert C. Moore CSLI–85–33 ($*2.00*)

Applicability of Indexed Grammars to Natural Languages Gerald Gazdar CSLI–85–34 ($*2.00*)

Commonsense Summer: Final Report Jerry R. Hobbs, et al CSLI–85–35 ($*12.00*)

Limits of Correctness in Computers Brian Cantwell Smith CSLI–85–36 ($*2.50*)

The Coherence of Incoherent Discourse Jerry R. Hobbs and Michael H. Agar CSLI–85–38 ($*2.50*)

A Complete, Type-free "Second-order" Logic and Its Philosophical Foundations Christopher Menzel CSLI–86–40 ($*4.50*)

Possible-world Semantics for Autoepistemic Logic Robert C. Moore CSLI–85–41 ($*2.00*)

Deduction with Many-Sorted Rewrite José Meseguer and Joseph A. Goguen CSLI–85–42 ($*1.50*)

On Some Formal Properties of Metarules Hans Uszkoreit and Stanley Peters CSLI–85–43 ($*1.50*)

Language, Mind, and Information John Perry CSLI–85–44 ($*2.00*)

Constraints on Order Hans Uszkoreit CSLI–86–46 ($*3.00*)

Linear Precedence in Discontinuous Constituents: Complex Fronting in German Hans Uszkoreit CSLI–86–47 ($*2.50*)

A Compilation of Papers on Unification-Based Grammar Formalisms, Parts I and II Stuart M. Shieber, Fernando C.N. Pereira, Lauri Karttunen, and Martin Kay CSLI–86–48 ($*10.00*)

Noun-Phrase Interpretation Mats Rooth CSLI–86–51 ($*2.00*)

Noun Phrases, Generalized Quantifiers and Anaphora Jon Barwise CSLI–86–52 ($*2.50*)

Circumstantial Attitudes and Benevolent Cognition John Perry CSLI–86–53 ($*1.50*)

A Study in the Foundations of Programming Methodology: Specifications, Institutions, Charters and Parchments Joseph A. Goguen and R. M. Burstall CSLI–86–54 ($*2.50*)

Intentionality, Information, and Matter Ivan Blair CSLI–86–56 ($*3.00*)

Computer Aids for Comparative Dictionaries Mark Johnson CSLI–86–58 ($*2.00*)

A Sheaf-Theoretic Model of Concurrency Luís F. Monteiro and Fernando C. N. Pereira CSLI–86–62 ($*3.00*)

Tarski on Truth and Logical Consequence John Etchemendy CSLI–86–64 ($*3.50*)

Categorial Unification Grammars Hans Uszkoreit CSLI–86–66 ($*2.50*)

Generalized Quantifiers and Plurals Godehard Link CSLI–86–67 ($*2.00*)

Radical Lexicalism Lauri Karttunen CSLI–86–68 ($*2.50*)

What is Intention? Michael E. Bratman CSLI–86–69 ($*2.00*)

The Correspondence Continuum Brian Cantwell Smith CSLI–87–71 ($*4.00*)

The Role of Propositional Objects of Belief in Action David J. Israel CSLI–87–72 ($*2.50*)

Two Replies Jon Barwise CSLI–87–74 ($*3.00*)

Semantics of Clocks Brian Cantwell Smith CSLI–87–75 ($*2.50*)

The Parts of Perception Alexander Pentland CSLI–87–77 ($*4.00*)

The Situated Processing of Situated Language Susan Stucky CSLI–87–80 ($*1.50*)

Muir: A Tool for Language Design Terry Winograd CSLI–87–81 ($*2.50*)

Final Algebras, Cosemicomputable Algebras, and Degrees of Unsolvability Lawrence S. Moss, José Meseguer, and Joseph A. Goguen CSLI–87–82 ($*3.00*)

The Synthesis of Digital Machines with Provable Epistemic Properties Stanley J. Rosenschein and Leslie Pack Kaelbling CSLI–87–83 ($*3.50*)

An Architecture for Intelligent Reactive Systems Leslie Pack Kaelbling CSLI–87–85 ($*2.00*)

Modular Algebraic Specification of Some Basic Geometrical Constructions Joseph A. Goguen CSLI–87–87 ($*2.50*)

Persistence, Intention and Commitment Phil Cohen and Hector Levesque CSLI–87–88 ($*3.50*)

Rational Interaction as the Basis for Communication Phil Cohen and Hector Levesque CSLI–87–89 ($*4.00*)

Models and Equality for Logical Programming Joseph A. Goguen and José Meseguer CSLI–87–91 ($*3.00*)

Order-Sorted Algebra Solves the Constructor-Selector, Mulitple Representation and Coercion Problems Joseph A. Goguen and José Meseguer CSLI–87–92 ($*2.00*)

Extensions and Foundations for Object-Oriented Programming Joseph A. Goguen and José Meseguer CSLI–87–93 ($*3.50*)

L3 Reference Manual: Version 2.19 William Poser CSLI–87–94 ($*2.50*)

Change, Process and Events Carol E. Cleland CSLI–88–95 ($*4.00*)

One, None, a Hundred Thousand Specification Languages Joseph A. Goguen CSLI–87–96 ($*2.00*)

Constituent Coordination in HPSG Derek Proudian and David Goddeau CSLI–87–97 ($*1.50*)

A Language/Action Perspective on the Design of Cooperative Work Terry Winograd CSLI–87–98 ($*2.50*)

Implicature and Definite Reference Jerry R. Hobbs CSLI–87–99 ($*1.50*)

Situation Semantics and Semantic Interpretation in Constraint-based Grammars Per-Kristian Halvorsen CSLI–87–101 ($*1.50*)

Category Structures Gerald Gazdar, Geoffrey K. Pullum, Robert Carpenter, Ewan Klein, Thomas E. Hukari, Robert D. Levine CSLI–87–102 ($*3.00*)

Cognitive Theories of Emotion Ronald Alan Nash CSLI–87–103 ($*2.50*)

Toward an Architecture for Resource-bounded Agents Martha E. Pollack, David J. Israel, and Michael E. Bratman CSLI–87–104 ($*2.00*)

On the Relation Between Default and Autoepistemic Logic Kurt Konolige CSLI–87–105 ($*3.00*)

Three Responses to Situation Theory Terry Winograd CSLI–87–106 ($*2.50*)

Subjects and Complements in HPSG Robert Borsley CSLI–87–107 ($*2.50*)

Tools for Morphological Analysis Mary Dalrymple, Ronald M. Kaplan, Lauri Karttunen, Kimmo Koskenniemi, Sami Shaio, Michael Wescoat CSLI–87–108 ($*10.00*)

Fourth Year Report of the Situated Language Research Program CSLI–87–111 (*free*)

Events and "Logical Form" Stephen Neale CSLI–88–113 ($*2.00*)

Backwards Anaphora and Discourse Structure: Some Considerations Peter Sells CSLI–87–114 ($*2.50*)

Toward a Linking Theory of Relation Changing Rules in LFG Lori Levin CSLI–87–115 ($*4.00*)

Fuzzy Logic L. A. Zadeh CSLI–88–116 ($*2.50*)

Dispositional Logic and Commonsense Reasoning L. A. Zadeh CSLI–88–117 ($*2.00*)

Intention and Personal Policies Michael Bratman CSLI–88–118 ($*2.00*)

Unification and Agreement Michael Barlow CSLI–88–120 ($*2.50*)

Extended Categorial Grammar Suson Yoo and Kiyong Lee CSLI–88–121 ($*4.00*)

Unaccusative Verbs in Dutch and the Syntax-Semantics Interface Annie Zaenen CSLI–88–123 ($*3.00*)

Types and Tokens in Linguistics Sylvain Bromberger CSLI–88–125 ($*3.00*)

Determination, Uniformity, and Relevance: Normative Criteria for Generalization and Reasoning by Analogy Todd Davies CSLI–88–126 (*$4.50*)

Modal Subordination and Pronominal Anaphora in Discourse Craige Roberts CSLI–88–127 (*$4.50*)

The Prince and the Phone Booth: Reporting Puzzling Beliefs Mark Crimmins and John Perry CSLI–88–128 (*$3.50*)

Set Values for Unification-Based Grammar Formalisms and Logic Programming William Rounds CSLI–88–129 (*$4.00*)

Fifth Year Report of the Situated Language Research Program CSLI–88–130 (*free*)

Locative Inversion in Chicheŵa: A Case Study of Factorization in Grammar Joan Bresnan and Jonni M. Kanerva CSLI–88–131 (*$5.00*)

An Information-Based Theory of Agreement Carl Pollard and Ivan A. Sag CSLI–88–132 (*$4.00*)

Relating Models of Polymorphism José Meseguer CSLI–88–133 (*$4.50*)

Psychology, Semantics, and Mental Events under Descriptions Peter Ludlow CSLI–89–135 (*$3.50*)

Mathematical Proofs of Computer System Correctness Jon Barwise CSLI–89–136 (*$3.50*)

The X-bar Theory of Phrase Structure András Kornai and Geoffrey K. Pullum CSLI–89–137 (*$4.00*)

Discourse Structure and Performance Efficiency in Interactive and Noninteractive Spoken Modalities Sharon L. Oviatt and Philip R. Cohen CSLI–90–138 (*$5.50*)

The Contributing Influence of Speech and Interaction on Some Aspects of Human Discourse Sharon L. Oviatt and Philip R. Cohen CSLI–90–139 (*$3.50*)

The Connectionist Construction of Concepts Adrian Cussins CSLI–90–140 (*$6.00*)

Sixth Year Report CSLI–90–141 (*free*)

Categorical Grammar Meets Unification Johan van Benthem CSLI–90–142 (*$4.50*)

Point of View Edit Doron CSLI–90–143 (*$3.50*)

Modal Logic as a Theory of Information Johan van Benthem CSLI–90–144 (*$5.50*)

What Is Information? David Israel and John Perry CSLI–91–145 (*$4.50*)

Fodor and Psychological Explanations John Perry and David Israel CSLI–91–146 (*$4.50*)

Decision Problems for Propositional Linear Logic Patrick Lincoln, John Mitchell, André Scedrov, and Natarajan Shankar CSLI–91–147 (*$10.00*)

Annual Report 1989-90 CSLI–91–148 (*free*)

Overloading Intentions for Efficient Practical Reasoning Martha E. Pollack CSLI–91–149 (*$5.50*)

Introduction to the Project on People, Computers, and Design Terry Winograd CSLI–91–150 (*$5.50*)

Ecological Psychology and Dewey's Theory of Perception Tom Burke CSLI–91–151 (*$3.50*)

The Language/Action Approach to the Design of Computer-Support for Cooperative Work Finn Kensing and Terry Winograd CSLI–91–152 (*$5.50*)

The Absorption Principle and E-Type Anaphora Jean Mark Gawron, John Nerbonne, and Stanley Peters CSLI–91–153 (*$6.00*)

Ellipsis and Higher-Order Unification Mary Dalrymple, Stuart M. Shieber, and Fernando C. N. Pereira CSLI–91–154 (*$5.50*)

Sheaf Semantics for Concurrent Interacting Objects Joseph A. Goguen CSLI–91–155 (*$5.00*)

Communication and Strategic Inference Prashant Parikh CSLI–91–156 (*$5.00*)

Shared Cooperative Activity Michael E. Bratman CSLI–91–157 (*$3.50*)

Practical Reasoning and Acceptance in a Context Michael E. Bratman CSLI–91–158 (*$3.50*)

Planning and the Stability of Intention Michael E. Bratman CSLI–91–159 (*$3.50*)

Logic and the Flow of Information Johan van Benthem CSLI–91–160 (*$5.00*)

Learning HCI Design: Mentoring Project Groups in a Course on Human-Computer Interaction Brad Hartfield, Terry Winograd, and John Bennett CSLI–91–161 (*$3.50*)

How to Read Winograd's & Flores's Understanding Computers and Cognitiion Hugh McGuire CSLI–92–162 (*$6.00*)

In Support of a Semantic Account of Resultatives Adele E. Goldberg CSLI–92–163 ($)

Augmenting Informativeness and Learnability of Items in Large Computer Networks Clarisse S. de Souza CSLI–92–164 ($)

Terry Winograd CSLI–92–165 ($)

A Semiotic Approach to User Interface Language Design Clarisse S. de Souza CSLI–92–166 ($)

Lecture Notes

The titles in this series are distributed by the University of Chicago Press and may be purchased in academic or university bookstores or ordered directly from the distributor: Order Department, 11030 S. Langely Avenue, Chicago, Illinois 60628.

A Manual of Intensional Logic. Johan van Benthem, second edition, revised and expanded. Lecture Notes No. 1. ISBN 0-937073-29-6 (paper), 0-937073-30-X (cloth)

Emotion and Focus. Helen Fay Nissenbaum. Lecture Notes No. 2. ISBN 0-937073-20-2 (paper)

Lectures on Contemporary Syntactic Theories. Peter Sells. Lecture Notes No. 3. ISBN 0-937073-14-8 (paper), 0-937073-13-X (cloth)

An Introduction to Unification-Based Approaches to Grammar. Stuart M. Shieber. Lecture Notes No. 4. ISBN 0-937073-00-8 (paper), 0-937073-01-6 (cloth)

The Semantics of Destructive Lisp. Ian A. Mason. Lecture Notes No. 5. ISBN 0-937073-06-7 (paper), 0-937073-05-9 (cloth)

An Essay on Facts. Ken Olson. Lecture Notes No. 6. ISBN 0-937073-08-3 (paper), 0-937073-05-9 (cloth)

Logics of Time and Computation. Robert Goldblatt, second edition, revised and expanded. Lecture Notes No. 7. ISBN 0-937073-94-6 (paper), 0-937073-93-8 (cloth)

Word Order and Constituent Structure in German. Hans Uszkoreit. Lecture Notes No. 8. ISBN 0-937073-10-5 (paper), 0-937073-09-1 (cloth)

Color and Color Perception: A Study in Anthropocentric Realism. David Russel Hilbert. Lecture Notes No. 9. ISBN 0-937073-16-4 (paper), 0-937073-15-6 (cloth)

Prolog and Natural-Language Analysis. Fernando C. N. Pereira and Stuart M. Shieber. Lecture Notes No. 10. ISBN 0-937073-18-0 (paper), 0-937073-17-2 (cloth)

Working Papers in Grammatical Theory and Discourse Structure: Interactions of Morphology, Syntax, and Discourse. M. Iida, S. Wechsler, and D. Zec (Eds.) with an Introduction by Joan Bresnan. Lecture Notes No. 11. ISBN 0-937073-04-0 (paper), 0-937073-25-3 (cloth)

Natural Language Processing in the 1980s: A Bibliography. Gerald Gazdar, Alex Franz, Karen Osborne, and Roger Evans. Lecture Notes No. 12. ISBN 0-937073-28-8 (paper), 0-937073-26-1 (cloth)

Information-Based Syntax and Semantics. Carl Pollard and Ivan Sag. Lecture Notes No. 13. ISBN 0-937073-24-5 (paper), 0-937073-23-7 (cloth)

Non-Well-Founded Sets. Peter Aczel. Lecture Notes No. 14. ISBN 0-937073-22-9 (paper), 0-937073-21-0 (cloth)

Partiality, Truth and Persistence. Tore Langholm. Lecture Notes No. 15. ISBN 0-937073-34-2 (paper), 0-937073-35-0 (cloth)

Attribute-Value Logic and the Theory of Grammar. Mark Johnson. Lecture Notes No. 16. ISBN 0-937073-36-9 (paper), 0-937073-37-7 (cloth)

The Situation in Logic. Jon Barwise. Lecture Notes No. 17. ISBN 0-937073-32-6 (paper), 0-937073-33-4 (cloth)

The Linguistics of Punctuation. Geoff Nunberg. Lecture Notes No. 18. ISBN 0-937073-46-6 (paper), 0-937073-47-4 (cloth)

Anaphora and Quantification in Situation Semantics. Jean Mark Gawron and Stanley Peters. Lecture Notes No. 19. ISBN 0-937073-48-4 (paper), 0-937073-49-0 (cloth)

Propositional Attitudes: The Role of Content in Logic, Language, and Mind. C. Anthony Anderson and Joseph Owens. Lecture Notes No. 20. ISBN 0-937073-50-4 (paper), 0-937073-51-2 (cloth)

Literature and Cognition. Jerry R. Hobbs. Lecture Notes No. 21. ISBN 0-937073-52-0 (paper), 0-937073-53-9 (cloth)

Situation Theory and Its Applications, Vol. 1. Robin Cooper, Kuniaki Mukai, and John Perry (Eds.). Lecture Notes No. 22. ISBN 0-937073-54-7 (paper), 0-937073-55-5 (cloth)

The Language of First-Order Logic (including the Macintosh program, Tarski's World). Jon Barwise and John Etchemendy, second edition, revised and expanded. Lecture Notes No. 23. ISBN 0-937073-74-1 (paper)

Lexical Matters. Ivan A. Sag and Anna Szabolcsi, editors. Lecture Notes No. 24. ISBN 0-937073-66-0 (paper), 0-937073-65-2 (cloth)

Tarski's World. Jon Barwise and John Etchemendy. Lecture Notes No. 25. ISBN 0-937073-67-9 (paper)

Situation Theory and Its Applications, Vol. 2. Jon Barwise, J. Mark Gawron, Gordon Plotkin, Syun Tutiya, editors. Lecture Notes No. 26. ISBN 0-937073-70-9 (paper), 0-937073-71-7 (cloth)

Literate Programming. Donald E. Knuth. Lecture Notes No. 27. ISBN 0-937073-80-6 (paper), 0-937073-81-4 (cloth)

Normalization, Cut-Elimination and the Theory of Proofs. A. M. Ungar. Lecture Notes No. 28. ISBN 0-937073-82-2 (paper), 0-937073-83-0 (cloth)

Lectures on Linear Logic. A. S. Troelstra. Lecture Notes No. 29. ISBN 0-937073-77-6 (paper), 0-937073-78-4 (cloth)

A Short Introduction to Modal Logic. Grigori Mints. Lecture Notes No. 30. ISBN 0-937073-75-X (paper), 0-937073-76-8 (cloth)

Other CSLI Titles Distributed by UCP

Agreement in Natural Language: Approaches, Theories, Descriptions. Michael Barlow and Charles A. Ferguson (Eds.). ISBN 0-937073-02-4 (cloth)

Papers from the Second International Workshop on Japanese Syntax. William J. Poser (Ed.). ISBN 0-937073-38-5 (paper), 0-937073-39-3 (cloth)

The Proceedings of the Seventh West Coast Conference on Formal Linguistics (WCCFL 7). ISBN 0-937073-40-7 (paper)

The Proceedings of the Eighth West Coast Conference on Formal Linguistics (WCCFL 8). ISBN 0-937073-45-8 (paper)

The Phonology-Syntax Connection. Sharon Inkelas and Draga Zec (Eds.) (co-published with The University of Chicago Press). ISBN 0-226-38100-5 (paper), 0-226-38101-3 (cloth)

The Proceedings of the Ninth West Coast Conference on Formal Linguistics (WCCFL 9). ISBN 0-937073-64-4 (paper)

Japanese/Korean Linguistics. Hajime Hoji (Ed.). ISBN 0-937073-57-1 (paper), 0-937073-56-3 (cloth)

Experiencer Subjects in South Asian Languages. Manindra K. Verma and K. P. Mohanan (Eds.). ISBN 0-937073-60-1 (paper), 0-937073-61-X (cloth)

Grammatical Relations: A Cross-Theoretical Perspective. Katarzyna Dziwirek, Patrick Farrell, Errapel Mejías Bikandi (Eds.). ISBN 0-937073-63-6 (paper), 0-937073-62-8 (cloth)

The Proceedings of the Tenth West Coast Conference on Formal Linguistics (WCCFL 10). ISBN 0-937073-79-2 (paper)

Books Distributed by CSLI

The Proceedings of the Third West Coast Conference on Formal Linguistics (WCCFL 3). (*$10.95*) ISBN 0-937073-45-8 (paper)

The Proceedings of the Fourth West Coast Conference on Formal Linguistics (WCCFL 4). (*$11.95*) ISBN 0-937073-45-8 (paper)

The Proceedings of the Fifth West Coast Conference on Formal Linguistics (WCCFL 5). (*$10.95*) ISBN 0-937073-45-8 (paper)

The Proceedings of the Sixth West Coast Conference on Formal Linguistics (WCCFL 6). (*$13.95*) ISBN 0-937073-45-8 (paper)

Hausar Yau Da Kullum: Intermediate and Advanced Lessons in Hausa Language and Culture. William R. Leben, Ahmadu Bello Zaria, Shekarau B. Maikafi, and Lawan Danladi Yalwa. (*$19.95*) ISBN 0-937073-68-7 (paper)

Hausar Yau Da Kullum Workbook. William R. Leben, Ahmadu Bello Zaria, Shekarau B. Maikafi, and Lawan Danladi Yalwa. (*$7.50*) ISBN 0-93703-69-5 (paper)

Ordering Titles Distributed by CSLI

Titles distributed by CSLI may be ordered directly from CSLI Publications, Ventura Hall, Stanford University, Stanford, California 94305-4115 or by phone (415)723-1712 or (415)723-1839. Orders can also be placed by e-mail (pubs@csli.stanford.edu) or FAX (415)723-0758.

All orders must be prepaid by check, VISA, or MasterCard (include card name, number, expiration date). For shipping and handling add $2.50 for first book and $0.75 for each additional book; $1.75 for the first report and $0.25 for each additional report. California residents add 7% sales tax.

For overseas shipping, add $4.50 for first book and $2.25 for each additional book; $2.25 for first report and $0.75 for each additional report. All payments must be made in US currency.

CSLI was founded early in 1983 by researchers from Stanford University, SRI International, and Xerox PARC to further research and development of integrated theories of language, information, and computation. CSLI headquarters and the publication offices are located at the Stanford site.

CSLI/SRI International
333 Ravenswood Avenue
Menlo Park, CA 94025

CSLI/Stanford
Ventura Hall
Stanford, CA 94305

CSLI/Xerox PARC
3333 Coyote Hill Road
Palo Alto, CA 94304

99 98 97 96 95 94 93 92   5 4 3 2 1

Library of Congress Cataloging-in-Publication Data

Goldblatt, Robert.
Logics of time and computation / Robert Goldblatt. — 2nd ed., rev. and expanded.
p. cm. — (CSLI lecture notes ; no. 7)
Includes bibliographical references and indexes.
ISBN 0-937073-93-8 — ISBN 0-937073-94-6 (pbk.)
1. Modality (Logic). I. Title. II. Series.
QA9.46.G65 1992
160–dc20   92-12978
CIP

CSLI Lecture Notes report new developments in the study of language, information, and computation. In addition to lecture notes, the series includes monographs, working papers, and conference proceedings. Our aim is to make new results, ideas, and approaches available as quickly as possible.